PRACTICAL EMULSIONS

PRACTICAL EMULSIONS

Volume 1

Materials and Equipment

H. Bennett
Technical Director
B.R.Laboratory
Miami Beach, Florida

Jack L. Bishop Jr., Ch.E.
Dow Corning Corporation
Midland, Michigan

Max F. Wulfinghoff, M.E.
Tampa, Florida

CHEMICAL PUBLISHING COMPANY, INC.
New York

ISBN 0-8206-0257-4

Third Edition

Printed in the United States of America

Preface To The Third Edition

The third edition has maintained the tone and some of the practical hints established by Mr. Bennett in the first two editions. The current theories concerning emulsions have been added to enable the up-to-date technologist to follow the literature in the field. The section on equipment has been expanded to cover auxiliary equipment as well as the standard emulsification equipment. As many disciplines are involved in the study of the theory and practice of emulsions, many types of sources were consulted to produce information of a scope to be of help to readers with diverse backgrounds. Some reference tables have been incorporated for the convenience of the reader. In addition, the bibliography has been brought up to date. Much assistance was given by companies whose products are mentioned in the book.

The authors wish to thank especially Mr. J. Cekada, Mr. D. E. Findlay, Mr. M. E. Nelson, and Dr. L. J. Tyler—all of Dow Corning Corporation—and Dr. E. Hodnett—formerly of Dow Corning, now in London—for their comments and aid. They also wish to thank Dow Corning Corporation for providing the time and secretarial assistance for Mr. Bishop to prepare his portion of this book.

February 1967 J. L. Bishop, Jr.

Contents

Introduction

Emulsions, though not new, are finding new and wider applications daily. One of the first references to emulsions was recorded by Galen (131-c.201), the Greek physician.[1] Beginning with that early reference to the emulsifying power of beeswax, the art and science of emulsification has flourished.

Emulsions are prepared and used for a variety of reasons. As oil paint cannot be applied to a damp surface, it is emulsified in water. The oil paint, then, in the form of an emulsion, can be applied to a damp surface. Therefore, the emulsion can change the application characteristics of a material.

Water is a desirable, cheap diluent, and an emulsion is an easy method of using water to dilute materials that are not soluble in water. In addition, the fire hazard of flammable water-insoluble materials can be decreased through emulsification.

The odor and taste of water-insoluble materials can be reduced by the use of an emulsion. Cod-liver oil, for example, loses much of its fishy, oily taste when it is emulsified.

The kinetics of many reactions are enhanced through the use of emulsion polymerization techniques.

On the other hand, emulsions are difficult to manufacture. A small deviation in temperature or mixing speed or small amounts of impurities can prevent the formation of a stable emulsion. Emulsions are sensitive in varying degrees to heat, cold, and age. The production of good, stable emulsions, therefore, is the combination of science and art. It is the purpose of this book to describe the art and technique of emulsification.

Reference

1. Becher, P., *Emulsions, Theory and Practice*, p. 95; Reinhold Publishing Corp., New York, 1965.

chapter 1

BASIC CONSIDERATIONS

SOLUTIONS, DISPERSIONS, SUSPENSIONS

Before discussing emulsions, the differences among emulsions, solution, dispersions, and suspensions should be explained. All of these categories may be considered somewhat overlapping subdivisions of fluid mixtures.

A *solution* is the most intimate type of mixture because it is a single-phase, homogeneous mixture. A solution is generally the mixture of two or more substances on the molecular or ionic level. The material present in lesser quantity in the solution is called the *solute*, and the material in greater concentration is usually called the *solvent*, or carrier.

Dispersions, suspensions, and emulsions are two-phase mixtures.

A *dispersion* is a physical, usually temporary, mixture of two mutually insoluble phases. A solid—the dispersed phase—is usually finely divided and mixed into a liquid—the dispersing medium phase. If the dispersion is of a more lasting nature, it is said to be a *suspension*. The difference between a dispersion and a suspension is not distinct. Generally, however, the smaller the particle size of the dispersed substance phase and the more nearly equal the specific gravities of the two phases, the more lasting will be the dispersion. A suspension of very small particle size (colloidal) may begin to approach the appearance, and have many of the attributes, of a solution.

A colloidal system can be further subdivided depending on the relationship between the internal (dispersed) and external (continuous) phases. If there is little attraction between the dispersed particles and the dispersing medium, the system may be called a *suspensoid* or a *lyophobic* (lack of strong affinity) colloid. If, however, there is a strong attraction between the particles and the dispersing medium, the system is called an *emulsoid* or a lyophilic (showing affinity) colloid. If a stabilizing agent (emulsifying agent) is required to provide a greater or lesser amount of stability to a lyophilic colloid, the colloid is called an *emulsion*.

Therefore, going from larger particle size and lowest stability to smaller particle size (molecular or ionic) and greatest stability, one can classify the mixtures as follows:

DISPERSION
SUSPENSION
General Suspension
Colloidal Suspension
Lyophobic Colloid
(Suspensoid)
Lyophilic Colloid
(Emulsoid)
EMULSION
SOLUTION

SUSPENSOIDS (LYOPHOBIC COLLOIDS)

A number of statements can be made about lyophobic colloids in general. Typically, the concentration of the internal phase and the molecular weights of the components of the suspensoid are low. Because the concentration of the internal phase is low, the viscosity of the suspensoid is similar to that of the dispersing medium.

As a result of molecular collisions between the dispersed particles and the dispersing medium, the suspensoid will exhibit the familiar Brownian Movement. As the forces involved in these collisions are greater than the force effect of the difference in specific gravity between the phases, the dispersed particles will remain suspended. In addition, as all of the dispersed particles have a similar electrical charge (either positive or negative), the particles repel one another. This repulsion in the dispersed phase reduces the likelihood of coalescence or settling and increases the stability of the suspensoid. If the amount of charge on a dispersed particle is high, however, coalescence can occur; particles with a +5 charge will tend to coalesce sooner than particles with a +2 charge.

Finally, suspensoids do not generally exhibit reversible dilution or concentration. Evaporation of the dispersing medium will tend to cause coalescence of the dispersed particles. Subsequent dilution will not return the suspensoid to the original state.

EMULSOIDS (LYOPHILIC COLLOIDS)

The concentration of the internal phase and its molecular weight may be higher in a lyophilic colloid than in a lyophobic colloid. Generally, an emulsoid will exhibit a somewhat higher viscosity than its continuous phase alone.

Brownian Movement is less noticeable in an emulsoid than in a suspensoid. As a result, the stability of an emulsoid is a function of the interaction between dispersed particle and dispersing medium. If the dispersed particles have a high degree of affinity for the dispersing medium, the latter will be attracted to dispersed particles. This attraction, in turn, results in the formation of a type of protective layer around each discrete dispersed particle. The protective layer tends to prevent coalescence of the dispersed particles.

The particulate electrical charge in an emulsoid is a function of the acidic or basic nature of the dispersant. If the dispersant is acidic, the emulsoid itself will be acidic, and vice-versa. Because the electrical attraction between particles is responsible for much of the stability of an emulsoid, the addition of a copious amount of an electrolyte will cause the dispersed phase to coagulate ("salt out").

An emulsoid, unlike a suspensoid, is reversible. Within practical limits, evaporation of the dispersant and dilution with the dispersant will have no effect on the stability of an emulsoid.

Because a protected (or stabilized) colloid acquires the properties of the stabilizing colloid, suspensoids (lyophobic colloids) can be stabilized by the addition of emulsoids (lyophilic colloids).

SURFACE ACTIVITY

The physical and chemical properties of a colloidal system are influenced by the small particle size of the disperse phase. The smaller the particle size in the dispersed phase, the larger the effective surface (surface area of unit volume) becomes. The physical properties of solubility, melting point, heat of solution, and color are affected by high effective surface area. In addition, chemical reactions, catalysis, adsorption, and heat-transfer are affected by the number of surface molecules (ions, atoms) to the volume of dispersed phase. Emulsions have a high relationship of surface area to volume, and the importance of emulsion polymerization, (to be discussed later) is a direct result of this high *surface area to volume* relationship.

EMULSIONS

An *emulsion* is usually a dispersion of two mutually insoluble liquids, e.g., oil and water. The term *emulsion* is broad enough to include lyophobic colloids and macro-colloidal dispersions. They range in range in stability from the "shake-well-before-using" pharmaceutical products of a few years ago to the present homogenized milk.

Emulsions usually contain emulsifying agents (surfactants, surface-active agents) to stabilize the dispersion of the two insoluble liquids The dispersed fluid is often called the *internal* or *discontinuous phase*, and the dispersing medium is called the *external* or *continuous phase*. Although the term *fluid* is used, particularly the internal phase may consist of waxes, asphalts, or other solids.

Emulsions have been historically considered as dispersions of oil and

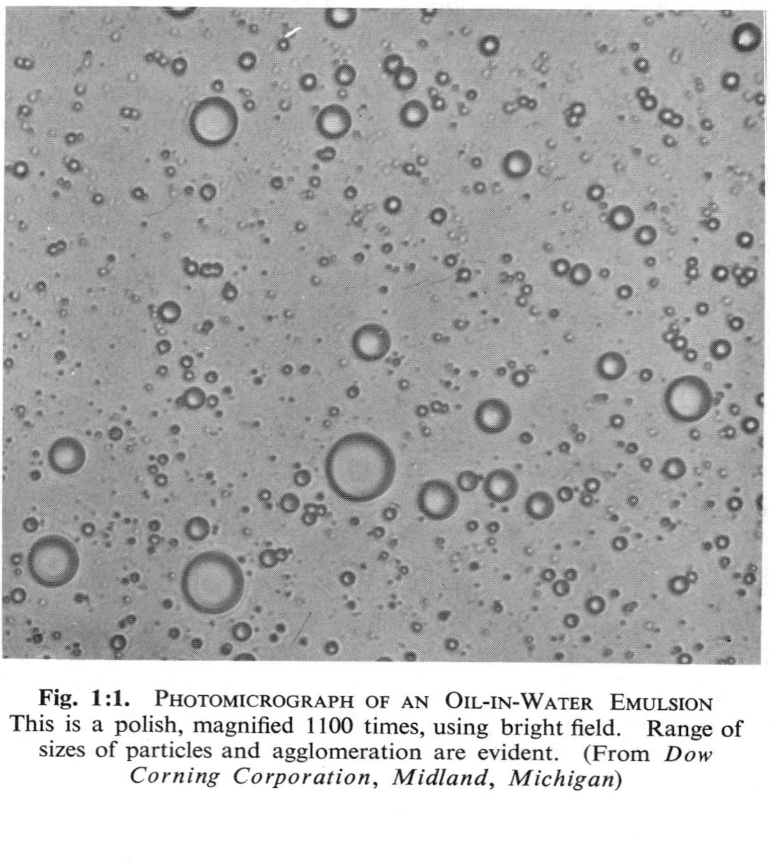

Fig. 1:1. PHOTOMICROGRAPH OF AN OIL-IN-WATER EMULSION
This is a polish, magnified 1100 times, using bright field. Range of sizes of particles and agglomeration are evident. (From *Dow Corning Corporation, Midland, Michigan*)

water. Even though one phase may be a solidified wax and the other a non-aqueous solvent, this simplified concept of an emulsion is encountered frequently. Within this simplified classification, emulsions are considered as *oil-in-water* emulsions, *water-in-oil* emulsions, or *dual* emulsions. The type of emulsion formed is a function of the viscosity, dielectric constant, relative concentration, and specific gravity of the two phases. The type of emulsifying agent and the pH of the the aqueous phase will also influence the type of emulsion.

Oil-in-Water Emulsion

An oil-in-water (O/W) emulsion consists of a dispersion with oil as the internal phase and water as the external or continuous phase. Systems containing more than 31% of water will generally form an oil-in-water dispersion. In some systems nearly 45% of water is necessary to form an oil-in-water emulsion. The oil-in-water type of emulsion is the most common and can be recognized by two simple tests. As water is the external (continuous) phase, a drop of an oil-in-water emulsion will spread when placed on the surface of water. This type of emulsion will be readily diluted by the addition of more water, but will not necessarily retain its stability. The addition of a water-soluble dye, such as food color, will immediately tint the emulsion.

If an oil-in-water emulsion is desired, the oil is generally added to the water. If the water is added to the oil, the emulsion will begin as a water-in-oil emulsion and gradually thicken as the water is added. A "thick phase" will result, which will quickly disappear with the addition of more water or mechanical agitation as the emulsion "inverts". This inversion marks the change of the water's from the internal to the external phase.

The use of *hydrophilic* (water-loving) emulsifying agents will also tend to form oil-in-water emulsions. The so-called *Hydrophile-Lipophile Balance* (HLB: see Chapter 8) of hydrophilic emulsifying agents will be greater than 11. Such emulsifying agents as gum arabic, gelatin, and soap tend to form oil-in-water emulsions.

Water-in-Oil Emulsion

A water-in-oil (W/O) emulsion consists of a dispersion that has oil as the external phase and water as the internal phase. Systems that contain less than 25% of water will generally form a water-in-oil emulsion. Sometimes less than 10% of water will be necessary to ensure a water-in-oil emulsion.

A water-in-oil emulsion can be recognized if it coalesces when placed on the surface of water; it cannot be diluted by water; and it takes up the color of an oil-soluble, rather than a water-soluble, dye. If a water-soluble food dye is added to a water-in-oil emulsion, the emulsion will not be tinted.

The addition of water to an oil, all other factors being equal, will favor the formation of a water-in-oil emulsion. Hydrophobic (lyophilic) emulsifying agents should be used to establish water-in-oil emulsions. These emulsifiers will have an HLB number below 9. Metallic soaps are a good example of a lyophilic emulsifying agent.

Inversion—Dual Emulsions

Inversion is the process of the simultaneous switching of the internal and external phases. This phase-reversal may be the result of mechanical or chemical action. If the emulsion contains 10—45% of water and is near an equilibrium between the oil-in-water and water-in-oil types, such a phase-reversal may occur as the result of pumping or other mechanical agitation. The chemical reaction between one of the phases and the emulsifying agent may chemically alter the latter sufficiently to permit phase-reversal. A change in *p*H or solubility of one or both of the phases over a period of time may also cause this phase-reversal. An emulsifying agent that is fairly soluble in both phases may cause phase-reversal if the relative concentration of the phases is altered.

An emulsion that is partially inverted is called a *dual emulsion*. A dual emulsion is usually the result of phase-reversal within the emulsion after it is formed. A portion of the continuous phase is present within the droplets of the dispersed phase in a dual emulsion.

The intentional preparation of a dual emulsion, or an emulsion that will invert upon standing, is to be avoided except in specialized instances. Unexpected inversion of the emulsion can completely negate the advantages of the initial emulsion. A cod-liver oil emulsion may leave the manufacturer as an aesthetically appealing product, only to invert to a smelly brew on the druggist's shelf.

A defoaming emulsion, however, is specifically designed to be inverted. This class of emulsions is stable on the shelf, but inverts to expose the defoaming oil upon dilution. Two types of dual emulsions can be illustrated. Lecithin dispersions form dual emulsions with benzene, carbon tetrachloride, chloroform, cresol, Hexalin, linseed oil, phenol, and light petroleum. Hevealipin, piptone, albumin, casein, and gelatin also form dual emulsions.

FOAMS

Foams may be considered emulsions in which the dispersed phase is a gas that is present in a fairly large proportion.

Kitchener and Cooper list three basic types of foams: *unstable*, *metastable*, and *chemical*.[1] Unstable foams collapse within a short time after they form, and the collapse is independant of drainage of the of the liquid phase from the gaseous bubbles. The drainage of film from a metastable foam eventually stops, but the foam is broken by exterior disturbing influences, e.g., vibrations, draft, evaporation, dust, and diffusion of gas.

Both unstable and metastable foams are similar to emulsions in that they are thermodynamically unstable (representing a higher level of energy). A chemical foam, on the other hand, is thermodynamically stable. One or more chemical reactions occur during or immediately after the formation of the foam to stabilize it. The popular automobile seats and mattresses of polyurethane foam are familiar examples of the chemical foam.

Pure liquids or saturated solutions do not produce foams.[2] Some impurity, no matter how dilute, is necessary before a system will foam. The most copious foams are produced by impurities which are slightly soluble (neither completely soluble nor completely insoluble in the liquid phase). Because surfactants, and particularly emulsifying agents, match this qualification so exactly, they are widely used in foams. A surfactant will tend to orient at any available interface because of its partial incompatibility with the liquid phase. The presence of a foam greatly enlarges the area of interfacial surface to which the emulsifying agent can migrate. Therefore, examination of the foam and its bulk liquid will find the surfactant concentrated in the foamed phase. This may result in relative starvation for surfactant in the body of the liquid. If the liquid is an emulsion itself, increased instability of the emulsion is inevitable.

Often protective colloids are added to foams. The protective colloid performs the same function in a foam that it does in a conventional emulsion. The colloid forms a protective layer at the interface of the dispersed phase (gas) to prevent droplets of gas from coalescing. The other discussions about conventional emulsions and their stability apply to foams as well.

The practical industrial uses of emulsifying agents and foams industrially include: fire fighting, shaving creams, beverages, ore-flotation and purification, and homemade cakes.

If a material is too soluble, or insoluble, it will not produce a copious foam. A 0.2% concentration of the dioctyl ester of sulfonated dicarboxylic acid forms more foam and a more stable foam than a 0.2% concentration of the dibutyl ester of sulfonated dicarboxylic acid. A 0.2% solution of a sulfated higher alcohol produces more of a froth and a more stable froth than either of the other two materials. This is in line with experience that detergent-type surface-active products are generally better foamers than wetting-agents.[3]

Organic acids, alcohols, amines, aldehydes, ketones, glycerol, and sugar aid in the production of foams but decrease their stability. The optimal concentration in thirty-five such compounds for a maximum of foam is given by Bartsch.[4]

Surface-activity, solubility, degree of dispersion, and viscosity are factors in the effectiveness of each substance. Molecular association or colloidal aggregation decreases the foam-building power of surface-active substances. High viscosity may not aid in the formation of foam, but it increases its stability. Mixed solutions of surface-active substances generally have greater foam-building power than the solution of just one component.

It is often necessary to break a foam (*de-foam*) or prevent one from forming (*anti-foam*). Oily layers on the surface of the liquid tend to prevent the formation of a foam because of the physical weight upon the foam. Other agents, such as the silicones, depress the surface tension and thus cause the collapse, or prevent the formation, of a foam. Whereas emulsifying agents are surface-active agents that are partially soluble in one or both phases of a system, de-foaming or anti-foaming agents are not compatible with either phase and may or may not have surface-active qualities.

Mechanical means (e.g., ultrasonics) are also used to de-foam systems. The prevention or destruction of a foam allows the surfactant in an emulsion to concentrate on the emulsification of the disperse phase. In the ore-flotation process foam must be created by a surfactant in order to float the mineral to the surface. After the foam is scraped off the surface, it must be broken to recover the mineral. The copious foaming of a shampoo is necessary for consumer acceptance, but the presence of this foam during the manufacture and packaging of the shampoo is inconvenient at best. The formation of foam in most industrial processes is undesirable because the foam occupies valuable space in equipment.

Each de-foaming or anti-foaming application often requires a specific agent. The amount of the agent that is used may range from 1 or 2% of some organic agents to 1 or 2 parts per million of the silicones. The subject of foam control will be discussed further in Chapter 3.

References

1. KITCHENER and COOPER, Foams, *Ind. Eng. Chem. Quarterly Review*, 13:71 (1959).
2. BICKERMAN, J. J., Foams and Emulsions:Formation, Properties Breakdowns, *Ind. Eng. Chem.*, 57:56, January 1965.
3. PORTER, *Ind. Eng. Chem., News Ed.*, 12:748 (1940).
4. SHERRICK, *Ind. Eng. Chem.*, 12:137 (1920).

chapter 2

PROPERTIES OF EMULSIONS

PARTICLE SIZE AND ARRANGEMENT

In the laboratory, the particle size of an emulsion is one of its most important characteristics. The particle size will determine the appearance of the emulsion and has a large effect on its stability. In order more easily to visualize the size of particles in an emulsion relative to natural objectives, Sutheim[1] developed comparisons in Table 2:1.

As the particle size in the dispersed phase of an emulsion decreases, its appearance changes from white to transparant. Correspondingly, the smaller particle sizes will usually favor greater stability.

The size of the particles in an emulsion is a function of the nature of the phases, the quantity and type of surfactant, and the processing techniques.[2]

A solid dispersed phase will form particles that are characteristic of the particular solid. The shape of the fluid particle, however, will be governed by the laws of thermodynamics. The most thermodynamically stable energy configuration in an emulsified droplet is spherical. This shape minimizes the surface area, and thus the surface energy, of the particle.

The same minimizing of surface energy dictates that the most stable configuration for the emulsion as a whole is two distinct phases (lowest thermodynamic energy level). Therefore, the type and amount of emulsifying agent must provide sufficient thermodynamic energy to maintain the dispersion of the particles of the internal phase. Increasing the amount of surfactant(s) in the emulsion will result in smaller particle size. Improving the hydrophilic-lipophilic balance will improve the efficiency of the surfactant(s) and decrease the particle size of the emulsion.

TABLE 2:1

RELATIVE PARTICLE SIZES[1]

Range of Dimensions	*Visibility*	*Description of State*	*Examples in Nature* (μ)		*Examples in Emulsions and Paints* (μ)	
10^{-1} cm(1 mm)	Plain visibility with the naked eye	—	Frog's egg	1000		
			Fine sand	500		
		Coarse dispersions				
10^{-2} cm			Ameba	100		
	Limit of visibility with the naked eye (approx. 50μ)	—	Potato starch	45-100	Mesh of No. 325 screen	44
					Coarse pigments	30
			Corn starch	15-20	Coarse emulsions	5-25
10^{-3} cm		Fine dispersions			Mesh of No. 1,250 screen	10
			Red blood corpuscle	7.5-8.5	Butterfat particles in milk	5-10
	Plain visibility with the microscope	—	Rice starch	3-7		
					Finest earth colors	2-3
10^{-4} cm(1μ)		Emusions	Average bacteria (typhoid)	1	Butterfat particles in homogenized milk	1-2
					Oil droplets in good emulsion vehicle	0.5-1.5
	Limit of microcopic resolving power (approx. 0.25 μ)	—	Wavelength of visible light	0.4-0.65	Precipited pigments	0.3-0.8
			Colloidal gold particles	0.2	Finest emulsion droplets	0.25
			Smallest bacteria	0.1	Lampblack	0.1
10^{-5} cm						
				(mμ)		(mμ)
		Colloid dispersions				
	Limit of ultramicroscopic resolving power (5-10 mμ)	—	Filterable virus (probable)	30		
10^{-6} cm					Colloidal carbon black	10-30
			Largest molecules, e.g., protein (probable)	6		
	Field of the eletron microscope	Molecular range				
			Palmitic acid (length of chain)	2.4		
10^{-7} cm(1 mμ)						
	Below any visibility		Methyl group	0.14		
10^{-8} cm(1Å)		Atomic range	Hydrogen (H_2)	0.1		

The mechanical energy put into the system during preparation may also provide the necessary thermodynamic energy and result in reduced particle sizes. Both the rate and duration of the agitation can be varied as well as the time when the emulsion is agitated. Through vigorous agitation at or near the inversion point, the effect of the mechanical mixing is amplified by the high viscosity of the emulsion. This "thick phase" technique will, therefore, reduce the particle size of some emulsions. Each system has an optimum surfactant and level of agitation.

W. C. Griffin summarizes the appearance of emulsions as a function of the particle size.[3] Their relationship is helpful to gain a qualitative estimate of emulsion particle size at a glance.

Particle Size	*Appearance*
Macroglobules	Two phases may be distinguished
Larger than 1μ	Milky white emulsion
1 to approx. 0.1μ	Blue-white emulsion
0.1 to 0.005μ	Gray semitransparent (translucent)
0.005μ and smaller	Transparent (translucent)

Quantitative means to determine particle size are discussed in Chapter 4. Briefly, however, the particle size distribution of an emulsion can be characterized by microscopy (optical, ultraviolet, electron), diffraction (x-ray or electron), dielectric constant, relaxation, etc.

RHEOLOGICAL PROPERTIES

Rheology is the study of the deformation and flow of matter. Knowledge of the rheological properties of an emulsion is important in the production and use of emulsions. The choice of mixer—type, size, optimum speed, and power requirements—is dependent upon the rheological properties of the emulsion. The ultimate use of the emulsion, in turn, dictates the rheological properties necessary for an acceptable product. For instance, a paint must flow easily onto a surface, yet it should not sag or run; a hand cream must be solid in the jar or tube, but spread easily over the hands.

Several terms are used to describe rheological phenomena. Whereas *fluidity* is a measure of the tendency of a fluid to flow, *viscosity* is a measure of its resistance to flow. Viscosity may be defined as the force necessary to compel two parallel liquid surfaces of unit area separated by a distance of one differential unit to slide past each other with a constant unit of velocity. Viscosity is more easily remembered as the factor of proportionality between rate of shear and shearing stress. The graphical representation of the functions of viscosity is called a *rheogram*. Six distinct types of rheology are defined in the literature and characterize the flow properties of fluids.

Newtonian Fluids

A Newtonian fluid is one in which the flow is independent of the amount of shear. Thus the rheogram will be a straight line, and flow is increased in direct proportion to the shearing stress.

Water is such a Newtonian fluid and has a viscosity of about one centipoise at 20°C. As the viscosity of all fluids is zero at zero shear stress, the rheogram of a Newtonian fluid is determined by only one viscosity measurement. The following fluids exhibit typical Newtonian behavior: glycerol, low-viscosity oils at low heat rates, and poly (dimethylsiloxane) fluids.

Non-Newtonian Fluids

Many fluids encountered in working with emulsions exhibit complex or non-Newtonian behavior. A non-Newtonian fluid is one in which the viscosity is a function of the amount of shear and the time of applying it. The behavior characteristic of non-Newtonian fluids places them in five major classes as follows: plastic, pseudoplastic, thixotropic, dilatant, and rheopectic.

Plastic:

The shearing stress of a *plastic* (or Bingham Plastic) fluid must exceed some minimum value before it flows. In practice, few materials exhibit this behavior, but modeling clay, cheese, and synthetic resins during molding[4] approach a plastic state.

Pseudoplastic:

Pseudoplastic behavior is frequently encountered. The consistency, or apparent viscosity, of a pseudoplastic substance decreases instantly with increasing rate of shear.[5] Dispersions of starch and methyl cellulose are common examples of pseudoplasticity.

Thixotropic:

Mayonnaise, paint, hand cream, and many other emulsions exhibit *thixotropic* behavior. The apparent viscosity in a thixotropic substance "decreases with time to some minimum value at any constant rate of shear."[6] Unlike pseudoplasticity, thixotropy is a reversible phenomenon.

Dilatant:

The opposite of a thixotropic substance is a *dilatant* material. Suspensions of starch in water and silicone bouncing putty (e.g., "Silly Putty") exhibit dilatant flow characteristics. The "apparent viscosity increases instantaneously with increasing rate of shear."[7] Dilatancy, like thixotropy, is a reversible flow property of a fluid.

Rheopectic:

The apparent viscosity of a *rheopectic* fluid "increases with time to some maximum value at any constant rate of shear."[8]

Rheology of Emulsions

The rheological properties of an emulsion are functions of six variables.[9]

1) The viscosity of the external phase
2) The viscosity of the internal phase
3) The relative volumes of the two phases (phase/volume ratio)
4) The nature of the emulsifying agent(s) and the interfacial film formed at the interface between phases
5) Electroviscous effect
6) Particle size distribution

An emulsion will generally have the rheological characteristics of the external phase. If the emulsion is at or near the point of inversion, the *phase/volume* ratio will become the determining factor in establishing the flow characteristics of the emulsion. Electro-viscous effects, such as in a highly ionic system, will encourage greater deviations from behavior as a Newtonian fluid. Furthermore, the addition of an electrolyte to a highly ionic emulsion will result in a dramatic decrease in in viscosity. The rheological behavior of an emulsion will be influenced by the internal phase in inverse proportion to the particle size of the internal phase. The type and characteristics of the interfacial film formed by the surfactant(s) influence the rheology of the entire system in proportion to its concentration.

Emulsions of high viscosity are often desirable. Cosmetic emulsions that can be applied in cream form, butter that does not flow, mayonnaise that can be spread with a knife, and a paint that will not run off the brush are examples of desirable high-viscosity emulsions. In butter, the solid fat is made the external phase of the emulsion so that the water is contained in a solid matrix (water-in-oil emulsion).

The other types of products mentioned may exhibit thixotropic behavior. Thixotropy implies a temporary gel-type structure which is easily ruptured by mechanical means to yield low-viscosity flow and later re-forms. A hand cream can be formulated near the inversion point so that the act of application will invert the emulsion and provide lotion-like spreading characteristics. The thixotropy built into a formulation for paint must rupture and re-form instantly to prevent dripping from the brush (high viscosity), ease of application (low viscosity) and non-sag on the vertical wall (high viscosity again). Many commercial agents (e.g., Thixins*) provide this effect through partial incompatibility with the external phase, weak chemical bonding (such as hydrogen bonding), or weak chemically or mechanically bound agglomerate chains.

The viscosity of emulsions may be increased by changing the ratio of the phases, increasing the amount of emulsifying agent, inverting the type of emulsion, and the addition of bodying agents. Such bodying agents for the O/W type of emulsions are hydrophilic colloids such as proteins, alginates, water-soluble gums, and water-soluble cellulose derivatives. High-viscosity emulsions tend to be stable because creaming and agglomeration of oil particles are greatly reduced. For the W/O emulsion, metallic soaps like aluminum stearate or calcium linoleate may be used as bodying agents.

Some practical pointers on the art of controlling the viscosity: of emulsions are as follows:

Irregularities in the measurement of viscosity of the same emulsion are most pronounced when the emulsion is freshly made. Exclusion of air from the emulsion gives more consistent viscosity measurements.[10]

The viscosity of gelatin emulsions is reduced by hydrolysis on aging, heating, or exposure to ultrasonic waves.[11]

The viscosity of natural W/O petroleum emulsions is increased by ferric ions.[12] In soap solutions the viscosity is lowered by small additions of electrolytes. This is due to loss of water by the colloidal particle and consequent shrinkage of the particle.

Mathematical correlations to establish viscosity have been presented by Clayton,[13] Taylor,[14] Monson,[15] Calderbank and Moo-Young,[16] Metzner,[17] and others. A collection of data on viscosities of frequently encountered materials is contained in the *International Critical Tables*.

Individual data on rheological properties should refer to the instrument and to the temperatures at which the measurements were taken. It must be possible to correlate such data to actual operating temperatures expected in the industrial process. Because, for instance, the viscosity of most liquids decreases for increasing temperatures, data on

* Baker Castor Oil Co.

viscosity that do not state the temperature are meaningless. Generally, if the temperature is not stated for data on viscosity, 25°C is understood.

MICELLE THEORY

In recent years several theories of emulsion technology have been based upon the concept of "micelles". A *micelle* is a small aggregation of surfactant molecules in the emulsion. These molecules of surfactant are oriented with the lipophilic end of the surfactant toward the oil phase and the hydrophilic end toward the aqueous phase.[18] The micelle theory concerns the surfactant and influences its behavior in the emulsion system. The geometric shape of these micelles is debatable and it may well be a function of the specific surfactant and its concentration.

The initial small amount of surfactant added to the bulk of system of pure water will solubilize and disperse. As more surfactant is added to the system, a dynamic equilibrium is formed between the loosely formed aggregates (micelles) and the solubilized surfactant. This equilibrium will prevail until the system is saturated with surfactant, at which point the formation of micelles becomes dominant. Further additions of surfactant will only form micelles.

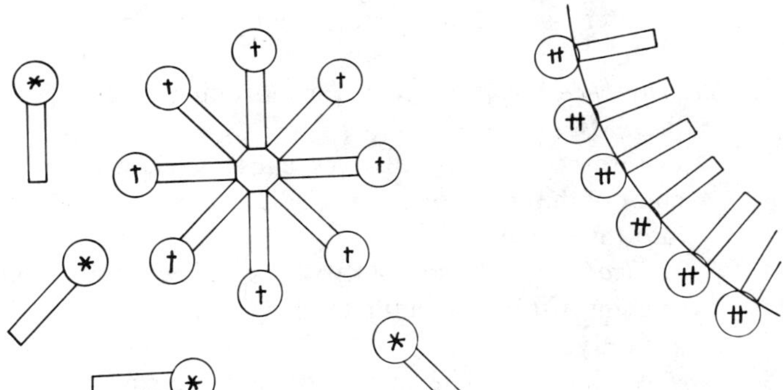

Fig. 2:1. Pictoral Representation of a Nonionic Surfactant. *, in solution; †, oriented in a micelle; ††, at surface of an emulsified particle.

This point of no further solubilization of added surfactant is physically observable through discontinuities or change in slope in the curves of surface tension, electrical conductance, etc., as a function of concentration of the surfactant. The quantity of micelles present in the system becomes appreciable at this point also. Accordingly, this point is called the *Critical Micelle Concentration* (CMC).

The practical effectiveness of surfactants as emulsifying agents only begins at the CMC. In many cases, detergency and foaming qualities are not appreciably improved after the CMC has been reached. For most surfactants, the CMC is generally less than 1 % of the total system.

The manner in which oil is incorporated into the micelle and the point at which (if ever) the micelle becomes a particle are open to debate. Many properties of surfactant solutions and emulsions are explained on the basis of this micelle theory. It may be remembered that the surfactant molecules, being only partially water-soluble, will tend to orient at any interface (air-water, water-container, water-oil) and the dynamic presence of micelles is but a manifestation of that thermodynamic tendency.

Shinoda *et al.*[19] thoroughly discuss the CMC and list twenty-one methods by which to measure it. The CMC is found by sharp discontinuity or change in slope of the following parameters as a function of a combination of surfactants:

1. *Electrical Conductance:* The electrical conductance of a water—surfactant system changes abruptly at the CMC.
2. *Transference Number:* The transport number changes at the CMC.
3. *Dye-Absorption:* Several dyes cause a change in the color of the emulsion at the CMC as they are absorbed into the micelles.
4. *Solubilization:* The solubility of a dye or organic liquid remains nearly constant to the CMC, then changes abruptly.
5. *Surface Tension:* The surface tension decreases rapidly until the CMC has been reached, then remains nearly constant with further additions of surfactant. This is the most common method of determining CMC in the laboratory.
6. *Partial Molal Volume:* The partial molal volume and density of the system change at the CMC.
7. *Refractive Index:* A change in the refractive index of the system occurs at the CMC.
8. *Light-Scattering:* The aggregation of surfactant molecules is reflected in the increase in scattered light.

9. *Freezing Point:* The CMC is related to the osmotic coefficient, but the applicability of determinations of the CMC at the solvent freezing point is small.
10. *Osmotic Pressure*
11. *Vapor Pressure:* The discontinuity in the depression of vapor pressure as a function of the surfactant concentration reveals the CMC.
12. *Solubility:* The solubility of a surfactant is a function of temperature and abruptly changes at the CMC.
13. *Electromotive Force:* The EMF can be used to observe changes in the CMC as a function of salt concentrations and other parameters.
14. *Sound Velocity:* The discontinuity in the velocity of sound through the emulsion as a function of surfactant concentration establishes the CMC.
15. *Polarographic:* The limiting diffusion current in the polarograph changes sharply at the CMC.
16. *Diffusion:* The rate of micellar diffusion is a function of the size and shape of the hydrated micelle. The discontinuity in the micellular diffusion as a function of surfactant concentration identifies the CMC.
17. *Viscosity*
18. *Dialysis*
19. *Electrophoretic Mobility:* A discontinuity in the migration as a function of surfactant concentration occurs at the CMC.
20. *UV and IR Absorption*: The absorption spectra undergo an abrupt change at the CMC.
21. *Depolarization of Fluorescence*: The depolarization of fluorescence of a suitable dye that has been solubilized by the micelle undergoes a change at the CMC.

Shinoda also discusses the effect of the molecular type of surfactant, blends of surfactants, and other parameters on the CMC.[20] Basically he found that critical micellar concentration is decreased with any of the following:

1. Increase in carbon atoms in the hydrocarbon.
2. Decrease in the number of ionic groups on the molecule.
3. Increase in the linearity of the molecule of the surfactant.
4. Position of the ionic group on the surfactant molecule

(CMC decreases as the ionic group is moved from the center to the end of the molecule).

5. Addition of a benzene ring to the oleophilic portion of the molecule.
6. Decreasing saturation in the molecule.
7. Removal of polar groups from the surfactant.
8. Substitution of $-CH_2-$ groups by $-CF_2-$ groups.
9. Increase in polyvalent character (bivalent groups have a smaller CMC than monovalent groups).

Mulley and Metcalf further discuss the Critical Micelle Concentration for a group of nonionic surface-active agents.[21]

In the selection of a surfactant to produce a desired emulsion, the material that produces the lowest Critical Micelle Concentration will—all other things being equal—tend to form a more stable emulsion. In addition, the surfactant that produces the smallest micelle will tend to produce the smallest emulsion particle and the most stable emulsion.

THE IMPORTANCE OF STABILITY

Stability is one of the key problems of the formulator or manufacturer of emulsions. An emulsified product loses a great deal of its appeal to the user if it separates into two layers, changes in consistency, discolors, or turns rancid before or while it is used. The consumer expects the cold cream to remain a cream and is not particularly concerned that the emulsion was frozen twice and heated to boiling before he used it. Thus the chemist must provide a margin of safety in formulating the emulsion so that it will reach the consumer in the desired form.

Dean[22] sums up some of the basic problems of stability in emulsions.

> Any emulsion is inherently unstable and will eventually break down to form two layers. Emulsions are unstable because the system has a large interfacial surface area and an interfacial surface energy which is proportional to this area. A change in the system which reduces the total surface energy will bring the system into a more stable state. The total surface energy is reduced if the emulsion drops coalesce, because surface energy is reduced when two drops join to make one drop. ... One way in which surface active agents promote the stability of emulsions is by reducing the total surface energy of the system.

INTERFACIAL PHENOMENA

The surfactant, as it orients at the oil-water interface, will form an interfacial film. The nature of this interfacial film is also important in determining the stability of the emulsion.[23]

Becher[24] pointed out that the strength and compactness of the interfacial films are important from a standpoint of stability. The interfacial film can have physical strength to reduce penetration from the continuous phase or other particles. The ionized electrical charge of the interfacial film may be strong enough to repel like particles. The addition of more surfactant will generally yield stronger interfacial films. Mixtures of emulsifying agents can also increase the strength of the interfacial film by allowing close packing of molecules at the interface. Other methods to increase the viscosity and elasticity of the interfacial film have also been discussed as means to improve the stability of an emulsion. If this film is of an ionic nature, electrical repulsion between like-charged particles is believed to enhance the stability of the emulsion.

Three basic theories concerning such charged particles have been advanced. If a molecule of a surfactant is oriented at the interface, there is oil on one side of the interface and water on the other. If a single layer of oriented, ionized molecules of surfactant comprises the interface, the electrical potential will decrease sharply and continuously into the bulk of the fluid. This situation is called a *Helmholtz Double Layer*. If, however, the ionized surfactant dissolved in the mass of fluid is also oriented to a greater or lesser extent, a *Gouy Diffuse Double Layer* is present. A further refinement by Stern hypothesizes a sharp Helmholtz layer of ionic thickness at the interface followed by a Gouy exponential electrical potential drop into the bulk of the fluid. In the emulsion, these effects are two-sided; that is, they operate on both sides of the liquid-liquid interface.[25]

The film-forming properties are not the same with all emulsifying agents. For example, under the same conditions, the film formed by sodium oleate is much more stable than that of sodium stearate. The stability of a sodium stearate emulsion is even inferior to that of oleic acid alone. To make a stable emulsion, an emulsifying agent must give thin interfacial films that are stable under the conditions to which the emulsion will be exposed.

Another factor influencing the stability of an emulsion is its viscosity. Viscous emulsions are generally more stable than mobile ones. The high viscosity of the external phase inhibits coalescence of the dispersed droplets of emulsion or any internal mass transfer in the emulsion.

Important factors in stability of emulsions are the proportion and solubility of the emulsifying agent in the phase, the stability of the emulsifying agent in water and air, its resistance to bacterial activity, to heat, to acids and salts, and its ability to form elastic and thin adhesive layers. The specific gravity of the dispersed phase and of the surrounding medium influences the stability of the emulsions. The greater the difference in specific gravity, the greater is the tendency for the emulsion to break, to "oil", or "cream".

Concentration and Phase-Volume Ratio

In general, the phase-volume ratio, temperature effects, and electrolytes can affect the stability of an emulsion. Phase-volume ratio means the ratio of the volume of the internal phase to that of the external phase. The relative volumes of the phases affect the type and stability of an emulsion; e.g., if in making margarine, the milk is run into the large volume of oil while stirring, a W/O type results.[26] When the oil is mixed into the milk, an O/W type forms. In emulsions made with soap, there is no particular limit to the amount of soap, but the amount of water must be sufficient to produce a proper hydrating effect.[27]

The best stability is observed when the volume of the internal phase is 40—60% of the emulsion. If the concentration of the internal phase falls far below a volume of 30% or goes much above 70%, the emulsion will break easily. If the internal phase constitutes much more than 70%, the dispersed droplets are packed so closely together that they may lose their spherical shape. The interfacial film will then lose its continuity and the unprotected droplets will coalesce, or the emulsion may invert.

Effects of Temperature

Many emulsions are broken by freezing. Freezing breaks them by destroying the membrane of emulsifying agent surrounding each globule, rather than by the mere freezing of the continuous phase.[28] Some emulsions can be frozen solid and thawed without breaking. Glycols can be added to an oil-in-water emulsion to improve its freeze-thaw stability. Supercooling without actual freezing is often possible, but if freezing occurs the emulsion will usually break. On freezing and thawing, with resultant breakdown, the surfactant may disappear from the water phase.[29]

Heat may produce profound chemical or physical changes which may break an emulsion. For example, the heating of emulsions made with hydrophilic agents usually increases the of size the globules. Agar-stabilized emulsions break at temperatures above the gelling point of dispersions of agar. Emulsions of Lecithin are less affected by heat. Nonionic emulsifying agents with higher cloud points generally produce emulsions that are more resistant to heat than are their homologs of lower cloud point.

Heat should be used in the preparation of an emulsion only to render fluid a very viscous or solid phase. If heat is necessary in manufacturing emulsions, the aqueous phase should be heated to about the same temperature as the oil phase to avoid chilling and solidifying the solid phase during mixing. Temperatures should never be higher than necessary.

ELECTROLYTES

The dispersed droplets of an emulsion can carry an electric charge. Since electric charges of the same kind (positive or negative) repel each other, the droplets are kept dispersed partly because of the repellant electric charges they carry. The electric charge is proportional to the surface area; the smaller the droplet, the larger is the relative surface area and the higher the relative charge.

The addition of electrolytes may act favorably or unfavorably on the stability of an emulsion. Electrolytes discharging the ionic protection of the internal phase will promote coalescence. Electrolytes may act upon a surfactant to destroy its emulsifying ability. This happens after the electrolyte induces coagulation (salting out) or change of the chemical nature of the emulsifying agent (forming a precipitate). On the other hand it is possible that small quantities of electrolytes increase the electrical potential of the ionized layer of the interfacial film. They may sometimes be helpful in stabilizing emulsions.[30] Electrolytes of higher valence generally are more destructive than materials of lower valence. Whereas addition of anionic electrolytes has little effect on an anionic emulsion, they will tend to destroy a cationic system and vice-versa. Electrolytes have less effect on nonionic and amphoteric emulsions.

A difference of electrical potential often exists between the dispersed and the continuous phase. A difference of about 0.06 volts affords a fairly stable emulsion. When it becomes zero, the emulsion breaks.

Given the general concepts of stability, as described, it is appropriate to discuss the major types of instability before considering the causes of the instability. The question of protecting the emulsion from biological attack will be deferred to Chapter 3. Suffice it to say that biological activity will usually destroy the surfactant and thus it essentially results in a deficiency of surfactant.

Creaming and Breaking

The two major effects of instability in emulsions are *creaming* and *breaking*. Both creaming and breaking are easily observed.

An emulsion is considered to *break*, if it separates into two immiscible phases. The phases may be re-dispersed upon agitation, but the dispersion is no longer really an emulsion. The best example of this type of instability is the medicine that bears the secondary label "Shake Well Before Using".

A lesser degree of breaking is called *oiling*. Whereas five breaking is generally clear-cut, oiling is insidious and it may slowly become worse and worse. This evidence of instability can generally be recognized by a small accumulation of oil at the interface of container, air, and liquid. It usually begins with small particles of oil floating on the surface of the emulsion.

Creaming is not de-emulsification, but rather the creation of two emulsions, one floating on top of the other. In one of the two emulsions thus formed the concentration of the dispersed phase is proportionately higher than in the other. If the density of the internal phase is lower than that of the external phase, the usual upward creaming will result. If the density of the internal phase is greater, downward creaming will be found.

Although a creamed emulsion has not broken, creaming is usually accompanied by larger particle size, and thus may indicate instability. The creaming phenomenon obeys Stoke's Law governing the rate of gravitational sedimentation. The rate of creaming, then, is directly proportional to the difference in densities between the phases and the square of the radius of the globules, and inversely proportional to the viscosity of the external phase.

Two empirical tests for creaming are optical observation and dipping a finger into the emulsion. As creaming is associated with stratification of different particle sizes and concentration in one layer of the dispersed phase, the optically evident difference in color is pronounced. If there has been some creaming in an O/W emulsion, some of the emulsion will stick to the finger instead of draining off quickly.[31] In a W/O emulsion, the converse will be seen.

Creaming can be accelerated and easily seen in the laboratory.

If a small amount of dye, soluble in the continuous phase, is added to the emulsion and the mixture is centrifuged, the color imparted will make the creaming easily visible. Creaming can be reduced by(1) better choice of emulsifying agents,(2) by an increase in the viscosity of the continuous phase, or(3) by the addition of solvent or other material to the dispersed phase in order to equalize the specific gravities of continuous and discontinuous phases.

Siragusa succinctly summarizes some of the theories concerning stability of emulsions.[32] Viscosity, although it probably contributes to stability, is probably not a *primary* cause of it. Some feel that stability is promoted by the formation of a hydrated colloid at the interface, or an adsorbed film that surrounds and protects the dispersed particles. Lowered surface tension is believed to promote stability in an emulsion, and, as also discussed previously, the packing and orientation of the surfactant at the interface.

Danielli, Pankhurst, and Riddiford published six practical observations on the stability of emulsions.[33]

1. There is a correlation between surface viscosity and emulsion stability for long-chain compounds, e.g., potassium soaps, alcohols, sodium alkyl sulfates, and alkyl trimethyl ammonium bromides.
2. Mixtures of alkanols, in an homologous series of 12 to 18 carbon atoms, with sodium alkyl sulfates show higher surface viscosity and stability in emulsions if chains of components are of the same length.
3. If the surfactant is an ionized soap, the interfacial viscosity does not seem to play a part in emulsion stabilization.
4. If the surfactant is water-soluble, but not ionized, the emulsion will be water-in-oil, and surface plasticity appears to be the stabilizing factor.
5. The stability of an emulsion is related to the strength of the surfactant film.
6. It seems that the mechanical properties of the surfactant film are more important than the wetting characteristics of the surfactant.

CAUSES OF UNSTABLE EMULSIONS

Based upon the preceding discussion of the theory of stability in emulsions, a short discussion of twelve common causes of instability is appropriate. Although the mechanism for the separation of phases in each case has not been fully investigated, the general factors causing separation can be discussed.

Wrong Emulsifying Agent

The selection of an improper emulsifying agent is a major cause of instability. No single emulsifying agent or combination is satisfactory for all needs. Although judicious blends of emulsifying agents often improve stability, Miller has pointed out that the blending of two non-ionic surfactants is occasionally detrimental to the stability of an emulsion.[34] For instance, he formed stable emulsions with E-10 and E-5 ethoxylates, but when these were combined with E-6, E-2 or E-1 ethoxylates, the stability of the resultant emulsion was markedly reduced. By changing the emulsifying agent(s), the type of emulsion may be changed (e.g. from oil-in-water to water-in-oil) and the stability thus improved.

Incompatibility

If the emulsifying agent reacts chemically with some portion of the system, this incompatibility will result in loss of stability in the emulsion. Generally, the emulsion may contain an ingredient that is chemically incompatible with the surfactant; e.g., borax in an emulsion containing gum acacia, or mineral acids in an emulsion containing albumen. Likewise, incompatibility will result from attempts to make an acid emulsion with a soap as the emulsifying agent. The mixing of anionic and cationic surfactants is an excellent example of incompatibility. The anionic and cationic materials neutralize each other and lose their surface-active characteristics.

Electrolytes

The presence of electrolytes, such as the salts of calcium, magnesium, aluminum, copper, iron, zinc, bismuth, and mercury often cause instability in emulsions. Although small amounts of these salts may be added after the emulsion is formed, the presence of an electrolyte during the preparation of the emulsion is usually a guarantee of later instability. In fact, the addition of large amounts of electrolytes to an emulsion is an excellent means of breaking it.

Unidentified Type

The failure to distinguish the type an emulsion can also lead to instability of the product. Attempts to add water to a water-in-oil emulsion or oil to an oil-in-water emulsion will be rewarded by quick separation. Some formulators assume that because an emulsion contains more oil than water a product is a water-in-oil emulsion, and proceed accordingly. The results of relying on such an untested assumption are often embarrassing to the formulator.

Improper Manufacturing

Improper procedure in manufacturing can also result in an unstable product. Generally, solid materials should be melted, or at least suspended, before the actual emulsification. The questions of adding *which* phase to *which* phase at *what* rate, with *how much* shear at *what* point and at *what* temperatures and *what* cooling rate, must be answered. Variations in procedure will, at best, raise questions as to the stability of the final product.

Inadequate Dispersion

The use of inadequate equipment or improper technique will result in an unstable emulsion because of incomplete dispersion. The popularity of highly efficient dispersing devices (e.g., homogenizer or colloid mill) is evidence of the importance of effective dispersion to the attainment of stable emulsions. A number of emulsions will separate on standing when simply mixed but will show excellent stability when mixed in a colloid mill or homogenizer.

Excessive Dispersion

Excessive dispersion, however, can also result in instability. With continued dispersion, the particle size becomes too small (and surface area, therefore, too large) and the surfactant will be insufficient to prevent coagulation or coalescence of the particles. In addition, the mechanical force of shearing may physically destroy the emulsifying agent (e.g., vegetable mucilages).

Temperature

The temperature at which an emulsion is formed can also be a cause of later instability. Because of the high amount of energy that a colloid mill or homogenizer, adds to the system, a rise in temperature and consequent instability of the emulsion, is common. Similarly, the rate and degree of cooling may affect the stability of the final product. Emulsions containing soap are likely to be sensitive to cold; those containing animal products are likely to be sensitive to heat. Few emulsions can withstand even localized boiling. Although it is generally felt that the slow cooling of a warm emulsion is desirable from the standpoint of stability, the converse has been demonstrated.[35]

Insufficient Viscosity

Many emulsions will separate because of *Insufficient Viscosity.* When the viscosity of an emulsion is too low, the increased movement of particles will cause coagulation or coalescence in the dispersed phase. The addition of a thickening agent to the external phase is the easiest method of increasing the viscosity of an emulsion. The phase-volume ratio can also be altered to increase viscosity.

Decomposition of Emulsifying Agent

Many of the emulsifying agents are biodegradable. As such, they must be adequately protected from biological attack before the product is used. The insufficiency, or lack, of a proper preservative can result in the decomposition of some or all of the surfactant(s). The resultant instability caused by the loss of surfactant is obvious.

Improper Packaging

Improper Packaging is another cause of an unstable product. If the package is inadequate the water may evaporate and the emulsion will "dry out." Drying-out of the emulsion can change the phase-volume relationship and result in instability. The electrical anode-cathode effects that result from the storage of a highly ionic emulsion in a conductive container may lead to instability. Plastic containers often become deformed because of absorption of oil from a water-in-oil emulsion often occurs. The stability of a product cannot be ascertained except by aging under actual conditions of storage in the container in which it will be marketed.

Improper Storage

Improper Storage Conditions can undo all of the chemist's care in formulating a stable emulsion. The problem of actual storage of the product must be approached from two directions. First the chemist must endeavor to formulate an emulsion that will have adequate stability regardless of obusive storage conditions. Second, the chemist. must inform the user which storage conditions to avoid. Many formulators will still allow themselves a margin for error. For instance, a formulation that will survive three freezing-and-thawing cycles with little change in stability will still be labeled PROTECT FROM FREEZING. Cynical product development chemists often refer to this as the inclusion of the "idiot factor."

SUMMARY

Unstable Emulsions may be caused by:

- Improper ratio of oil and water phases
- Incorrect choice or amount of emulsifier
- Impurities in oil or water or emulsifier
- Reaction between two or more components and/or container.
- Overheating
- Excessive evaporation
- Freezing
- Mechanical shock or vibration
- Improper balancing of densities
- Improper time or speed of mixing
- Absorption of gases or foreign particles
- Large particles in disperse phase

DE-EMULSIFICATION

In many instances, it is desirable to break a stable emulsion or formulate a semi-stable one. Butter is obtained by breaking the milk emulsion, and water-in-oil emulsions found in the processing of petroleum must be broken as the first stage in petroleum refining.

Agricultural sprays and defoaming or antifoaming agents must be formulated as semi-stable emulsions; these products must break upon contact or dilution to release the active ingredients.

The techniques for breaking emulsions may be grouped as *mechanical*, *chemical*, or *electrical*. Mechanical techniques of de-emulsification include high rates of shear, settling, addition of solvents, heating, distilling, centrifuging, filtration, and sonic or ultrasonic vibration. Acids or ion-exchange resins are commonly used to break emulsions chemically, and while high-voltage currents will break them electrically.[36,38]

Becher views de-emulsification as a two-stage process. As the droplets in the dispersed phase combine to form aggregates, the emulsion is seen to pass through a stage of *flocculation*. The emulsion finally breaks as the aggregates *coalesce* and form a single, larger drop.[38] The literature offers number of practical applications to illustrate the extent and methods of de-emulsification.

For some purposes a quick-breaking emulsion is desirable, as in agricultural sprays. The emulsion should break as soon as possible after it is applied to the plant, and thus release the active ingredient. As such quick-breaking emulsions are usually unstable, they are made up just before use. Quick-breaking emulsions usually contain an insufficient amount of emulsifying agent.[39]

Occasionally it is necessary to break emulsions, e.g., in the production of butter from milk by churning. Although some emulsions can be broken easily by heat, cold, electrolytes, and mechanical means, others are resistant to breaking and must be handled by other means. For instance, a substance can be added which disrupts the film of the emulsifying agent or tends to reverse the type of emulsion; e.g., calcium chloride, which inverts an emulsion made with egg yolk and breaks those made with saponin.[40]

An example of chemical breaking is the de-emulsifying of a gelatin-stabilized emulsion of benzene in water by the addition of sodium hydroxide solution. If the sodium hydroxide is added immediately after emulsification, the separation of benzene starts at once; whereas if the emulsion is left undisturbed for some time before the sodium hydroxide is added, the separation is inhibited for a certain period. This period of inhibition is shown to be a parabolic function of the age of the emulsion. The limiting value found for this period varies directly with the concentration of gelatin used as stabilizer over the range noted from 5 minutes for 0.1% to 75 minutes for 0.4% gelatin. It is suggested that after emulsifying, the gelatin slowly forms protecting layers until equilibrium between the gelatin sol and the adsorbed gelatin is reached.[41]

A small addition of an alcohol breaks an emulsion of ether and water. An emulsion stabilized by soap may be broken by adding acid. Formaldehyde breaks emulsions stabilized by proteins. Anhydrous sodium sulfate, ammonium sulfate, or calcium chloride often salts out emulsifying agents. If the emulsion is present only as a broad layer at the interface between the two phases, it is best to draw off the emulsion separately and break it. Generally, the emulsion to be broken in the laboratory is placed in a separatory funnel. The breaking agent is added and the emulsion is vigorously agitated, then allowed to stand before the lower phase is drawn off.

Mechanical means are often used to break oil-in-water emulsions. Milk is frothed by introducing inert gas and the cream is separated by a sieve.[42] Butter is separated by boiling milk at a low temperature *in vacuo*.[43] Cream is churned in thin layers exposed to vibrational effects.[44] Wool fat is recovered from emulsions by foaming.[45,46] Emulsions stabilized by proteins can be broken by shaking, heat, or chemical denaturants.[47] W/O emulsions can be broken by such mechanical means as gravity separation,[48] circulating against baffles,[49] or oscillating plates,[50] heating,[51] heating under pressure,[52] and similar procedures.[53] Tar emulsions are broken with finely divided inorganic powders,[54] petroleum emulsions are forced through infusorial earth wetted with oil,[55] and through sand, silica, alumina[56] or porous diaphragms.[57]

In the laboratory one of the following methods usually succeeds in breaking emulsions: Addition of excess solvent, addition of a solvent miscible with each of the two phases, destruction of the emulsifying agent, salting out, filtration, heating, freezing, electrolyzing, centrifuging. Filtration breaks emulsions stabilized by flocculent precipitates.[58]

Underground water often forms troublesome emulsions in petroleum fields. The stability of these emulsions varies considerably. The amount of water in these emulsions ranges from a trace to more than 60% (average 25%). The processes for breaking such emulsions include filtration, evaporation, high-voltage alternating current or direct currents,[60] centrifuging and chemical treatment.[61]

References

1. SUTHEIM, G. M. AND MATTIELLO, J. J., *Protective and Decorative Coatings*, Vol. IV, p. 282; John Wiley and Sons, Inc., New York, 1944.
2. Processing techniques are discussed in Chapter 5.
3. W. C. GRIFFIN, *Encyclopedia of Chemical Technology*. (2nd Ed.) Vol. 3, p. 124; John Wiley-Interscience, New York, 1965.
4. PERRY, J. H., editor, *Chemical Engineering Handbook*, p. 1197, McGraw Hill Book Co., Inc., New York, 1950.
5. PERRY, *loc. cit.*
6. PERRY, *op. cit.*, p. 1198.
7. PERRY, *loc. cit.*
8. PERRY, *op. cit.*, p. 1199.
9. SHERMAN, P., *Research* (London) 8:396 (1955).
10. NARAYANASWAMY, *J. Indian. Inst. Sci.*, 17A:75 (1934).
11. PORTER, *Ind. Eng. Chem., News Ed.*, 12:748 (1940).
12. SHERRICK, *Ind. Eng. Chem.*, 12:137 (1920).
13. CLAYTON, *Theory of Emulsions*, Blakiston, McGraw-Hill, New York, 1935.
14. TAYLOR, *Treatise on Physical Chemistry*, D. Van Nostrand Co., Inc., Princeton, N.J., 1924.
15. MONSON, *Ind. Eng. Chem.*, 30:1287 (1938).
16. CALDERBANK and MOO-YOUNG, *Trans. Inst. Chem. Engrs.*, (London), 37 (1959).
17. METZNER, *J. AIChE* 7 (1961).
18. OSIPOW, L. I., *Surface Chemistry*, p. 164; Reinhold Publishing

Corp., New York, 1962.
19. SHINODA, K., NAKAGAWA, T., TAMAMUSHI, B., ISEMURA, T., *Colloidal Surfactants*, pp. 9-15 Academic Press, New York, 1963.
20. SHINODA, NAKAGAWA, TAMAMUSHI, ISEMURA, *op. cit.*, pp. 42-57.
21. MULLEY, B. A. and METCALF, A. D., Nonionic Surface Active Agents, *J. Colloid Sci.*, 17:523, August 1962.
22. DEAN, R. B., *Modern Colloids*, p. 227; D. Van Nostrand Co., Inc., Princeton, N., J., 1948.
23. KING, *Trans. Faraday Soc.*, 37:168 (1941).
24. BECHER, P., *Emulsions, Theory and Practice*, pp. 104-149; Reinhold Publishing Corporation, New York, 1965.
25. BECHER, *op. cit.*, pp. 104-149.
26. CLAYTON, *Margarine J.*, 2(30):503 (1920).
27. BROWN, *Textile Colorist*, 57:516 (1935).
28. ROCHOW, *Ind. Eng. Chem.*, 28:1926 (1936).
29. BENNETT, *Technical Aspect of Emulsions:* p. 146; Chemical Publishing Co., New York, 1936.
30. STEIK, *Am. Dyestuff Reptr.*, 25:429 (1936).
31. BROWN, *Textile Colorist*, 57:515 (1935).
32. SIRAGUSA, J. M., A Study of Some Emulsifiers for Pharmaceutical Emulsions: A doctoral dissertation; University of Florida, August 1955.
33. DANIELLI, J. F., Pankhurst, K. G. A., Riddiford, A. C., Editors, *Recent Progress in Surface Science;* Academic Press, New York, 1964.
34. MILLER, A., Emulsion Stability, *Drug and Cosmetic Ind.*, 97:679 (1965).
35. Ovide-Cool Emulsions, *Drug and Cosmetic Ind.*, 94: 759 (1964).
36. BECHER, P., *op. cit.* (24), p. 367.
37. MOLLIET, J. L., Collie, B., and Black, W., *Surface Activity*, (2nd Ed.) p. 264; D. Van Nostrand Co., Inc., Princeton, N.J., 1961.
38. BECHER, *op. cit.*, pp. 170-208.
39. BENNETT, *op. cit.* (29), p. 73.
40. BERKMAN, S., and EGLOFF, G., *Emulsions and Foams*, p. 4; Reinhold Publishing Corp., New York, 1941.
41. NUGENT, *Trans. Faraday Soc.*, October 27, 1921.
42. German Patent 314090 (1918).
43. British Patent 333233 (1929).
44. PERKS, British Patent 322416 (1929); US Patent 1791000 (1931).
45. BROADBRIDGE, US Patent 1505944 (1924); KARPINSKY, British Patent 177498 (1922); BAILEY, US Patent 1770476 (1930).
46. British Patent 328606 (1928); US Patent 1853871 (1932); German Patent 546231 (1929).
47. US Patent 1728011 (1930).
48. US Patents: BARBER, 1945766 (1934); EDWARDS, 1968131 (1934); MAHONE, 1940726 (1933); PINK, 1698002, 1698067 (1929).
49. FISHER, US Patent 1942794 (1933).

50. PALMER, US Patent 1781076 (1930).
51. US Patents: NOVOTNEY, 1968614 (1934); TURNER, 1948481 (1934); BRADY, 1754079 (1930); MOSEICKI, 1710374 (1929).
52. US Patents: CHAMPION, 1943367-8 (1934); DUDDS, 1123502 (1915); ELLIOTT, 1928282 (1933).
53. British Patent 26228 (1908).
54. ROBINSON, British Patent 268547 (1926); ROTH, US Patent 1665189 (1928).
55. ZOUL, US Patent 1501877 (1924).
56. US Patents: VAN LOENEN, 1944479-80 (1934); 1967601 (1934); GARRISON, 1947709 (1934); Eddy, 1807833 (1931); Pollock, 1847413 (1932).
57. US Patents: TRUMBLE, 1304124 (1919); HASEMAN, 1914665 (1933).
58. WEINBERGER, W., *Am. J. Pharmacy*, 102:594 (1930).
59. SHERRICK, *Ind. Eng. Chem.*, 12:133 (1920).
60. COTTRELL, US Patent 987114 (1911).
61. SHERRICK, *J. Chem. Soc.*, 67:63 (1895).

chapter 3

INGREDIENTS AND ADDITIVES

SURFACTANTS

A surfactant (surface-active agent) is a compound that lowers the surface tension and/or interfacial tension of an emulsion. This lowered interfacial tension reduces the cohesion in the internal phase. By thus hampering the coalescence of the dispersed phase, the stability of an emulsion or dispersion is improved.

Molliet, Collie, and Black[1] described a surfactant as a molecule with one or more groups that have an affinity for water, although one or more groups, usually at some separation from the former groups, are antipathetic to the medium. Thus, a surfactant will tend to orient at any and all interfaces if it is in a system in which it is partially soluble. They also state that adsorption is the fundamental phenomenon of surface activity. The adsorption of a surfactant leads to the lowering of one or more of the boundary tensions of the system and/or the stabilization of one or more of the interfaces in the system.[2] They also describe the surfactant molecule as being composed of a long-chain compound containing a small, oil-soluble

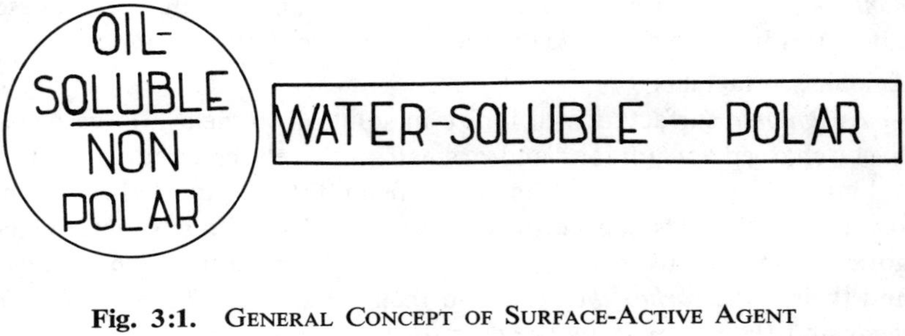

Fig. 3:1. General Concept of Surface-Active Agent

"head" and a large, water-soluble "tail" (or vice-versa).

The hydrophilic or water-loving groups on a surfactant molecule are usually characterized by polarity. Glycols, alcohols of lower molecular weight, and polyethers characterize the hydrophilic portion of a surfactant.

Schwartz and Perry[3] list seven hydrophobic groups that characterize the lipophilic, or non-polar, portion of the molecule.

1. Straight-chain alkyl groups of eight to eighteen carbon atoms derived from natural fatty acids.
2. Lower alkyl groups of three to eight carbon atoms that are frequently attached to an aromatic nucleus.
3. Branched-chain mono-olefins of eight to twenty carbon atoms that may be condensed with benzene also.
4. Modified petroleum hydrocarbons of eight to over-twenty carbon atoms, derived from kerosine, light oil, or the paraffin wax fraction.
5. Naphthenic acids from some types of petroleum.
6. Higher alcohols and hydrocarbons derived from Fischer-Tropsch and related syntheses.
7. Aromatics modified by terpenes or terpene-alcohols.

TYPES OF SURFACTANTS

Surfactants may be classified as *emulsifying agents*, *penetrants and wetting agents*, *dispersing agents*, *foaming agents*, and *protective colloids*. The boundaries of these categories often depend not upon the particular surfactant, but on its condition of use. This book will classify materials as emulsifying agents if they are often used for that purpose. In the other categories we shall discuss only materials specifically designed for each job.

The surfactants are chemically categorized as *anionic*, *cationic*, *nonionic*, *amphoteric*, *naturally occurring*, or oil-soluble. These categories indicate the ionic character of the molecule in solution.

Anionic Surfactants

An *anionic* surfactant will ionize in solution to yield a large negatively charged ion (anion) and a small positively charged ion. The anionic surfactant derives its name from the surface-active anion. Anionic surfactants can be further subdivided into four general categories based upon chemical structure: *sulfates*, *alkyl sulfonic acids* and their salts, *carboxylic acids* and their salts, and *alkylaryl sulfonic acids* and their salts (plus the ever-present *miscellaneous*).

Sulfated alcohols include the sodium or calcium salts of natural

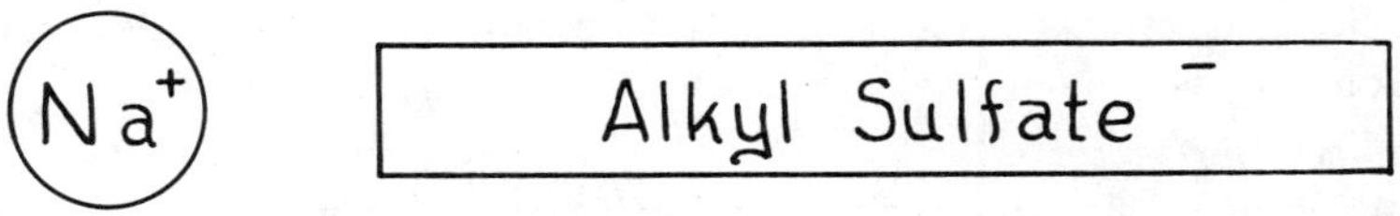

Fig. 3:2. SCHEME OF ANIONIC SURFACTANT

fats, e.g., coconut oil, palm-kernel oil, hydrogenated tallow, butyl alcohol, hydrobutyl alcohol, napthenyl alcohol, and synthetic alcohols. Sometimes the alcohols are modified before sulfonation by condensing the alcohol with some ethylene oxide to produce alkyl polyoxyethylene sulfates. Olefins are also sulfonated to yield a nonterminal sulfate group. The calcium salts have the advantage of forming water-soluble compounds in hard water so that they do not precipitate out of solution.[4] This general class of anionic surfactants is likely to hydrolyze.[5]

The alkyl sulfonic acids are often the by-product of petroleum refining (petroleum sulfonates) or from the addition of sulfur dioxide and chlorine to yield alkane sulfonyl chlorides. Other sulfonates are produced by adding sulfoacetic acid and either isethionic acid or *n*-methyl taurine to a fatty alcohol or oleic acid. Condensation products with ethylene oxide yield the ethoxylated alkyl sulfonates. Because they contain no ester or amide linkages, these materials are considered nonhydrolyzable.[6]

Both sodium and amine soaps of the carboxylic acids are *calcium-sensitive*. These materials form insoluble calcium salts in hard water. The sodium soaps of fatty acids such as stearic and lauric commonly have a *p*H of approximately 10. The amine soaps are generally the salts of substituted amines such as morpholine and triethanolamine. These soaps have a *p*H of approximately 8 and are usually formed *in situ*. Other sources of these anionic surfactants are rosin, tall oil, and napthenic acids. These carboxylic acid materials are both reasonably efficient and cheap emulsifying agents.[7]

The alkylaryl sulfonates are popular in the formulation of detergents. These materials consist of alkyl side-chain directly attached to a sulfonated aromatic group. A highly branched, alkyl side-chain attached to a sulfonated benzene ring yields the familiar ABS (alkyl benzene sulfonate) products. Alkylated naphthalenic sulfonates are also common members of this group.[8]

Miscellaneous anionic surfactants include phosphate and phosphoric acid esters, persulfates and thiosulfates, sulfonamides, and sulfonic acids.

Cationic Surfactants

A *cationic* emulsifying agent will ionize to form a large positively charged ion (cation) and a small negatively charged ion. Cationic

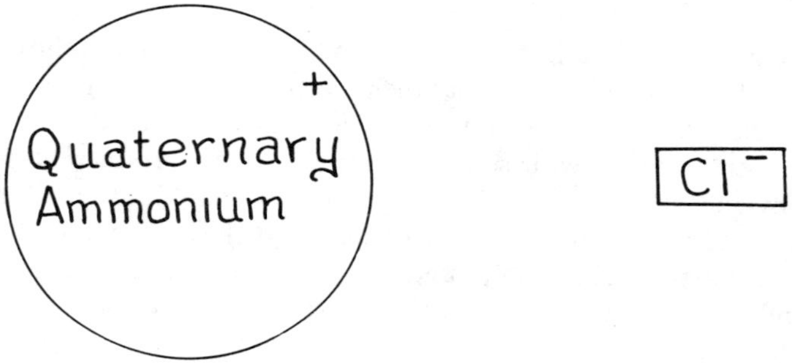

Fig. 3:3. SCHEME OF CATIONIC SURFACTANT

surfactants are usually characterized by the following three attributes:[9]

1. They are electrically attracted to, and bound by, surfaces with a negative charge, e.g., cellulose, proteins, wool, silk, and hair.
2. They give a soft feeling (or good "hand") to textiles, hair, and skin.
3. They have a high degree of bactericidal and fungicidal action.

These surfactants can be subdivided into *primary, secondary*, or *tertiary amine salts*, *quaternary ammonium compounds*, and *miscellaneous*. The last category includes other nitrogeneous bases of either quaternary or non-quaternary form (guanidine, thiuronium salts, etc.), and non-nitrogeneous bases (phosphonium or sulfonium compounds). Often the best emulsification is achieved by mixing cationics of varying molecular weights.[10]

The primary, secondary, or tertiary amine salts are generally too little soluble for wide use. As with other cationics, these salts are generally effective in acidic systems. Especially the acetateamine salts find some use at a *p*H less than 7.[11]

Quaternary ammonium compounds are fairly soluble in both acidic and basic systems, and many have some bactericidal activity. These compounds include alkyl halide or sulfate additions to tertiary amines, and dialkyl ethylenediamine and fatty acid chloride reaction products.[12]

Nonionic Surfactants

A *nonionic* surfactant will not ionize in solution. Nonionic surfactants are compatible with either anionic or cationic surfactants, and their effectiveness is fairly independent of hard water or the *p*H of the emulsion. The nonionic surfactants can be classified as

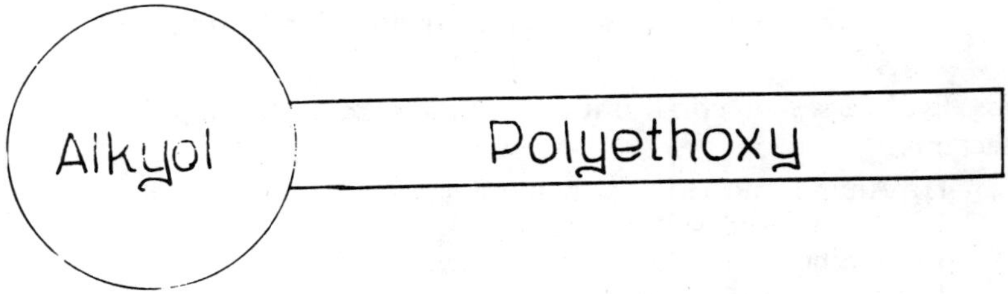

Fig. 3:4. SCHEME OF NONIONIC SURFACTANT

having alkoxy ether, alkoxy ester, or amide linkages to the solubilizing groups.

The ether-type materials are the result of condensing ethylene or propylene oxide upon a hydrophobic alcohol or mercaptan. The surface-active properties of the molecule can be altered by the amount and type of alkoxy compound added. Propylene oxide is also polymerized, and then ethylene oxide is added to one or both of the reactive groups on the newly formed basic molecule. This material is a fairly water-soluble block copolymer with useful surface-active properties.[13]

The reaction product of a fatty acid and a polyhydric alcohol is a useful alkoxy-ester nonionic surfactant. The dehydration of sorbitol yields the series of nonionic ester surfactants.[14]

The alkylolamine derivatives are an example of the amide type of nonionic emulsifying agents. Unsubstituted amides are also reacted with ethylene oxide to form ether-alcohol amides.[15]

Amphoteric Surfactants

Amphoteric (or *ampholytic*) surfactants assume either an anionic or cationic character. The ionic character of an amphoterc surfactant is a function of the *p*H of the system. Becher lists the following five classes of condensation products that are amphoteric surfactants:[16]

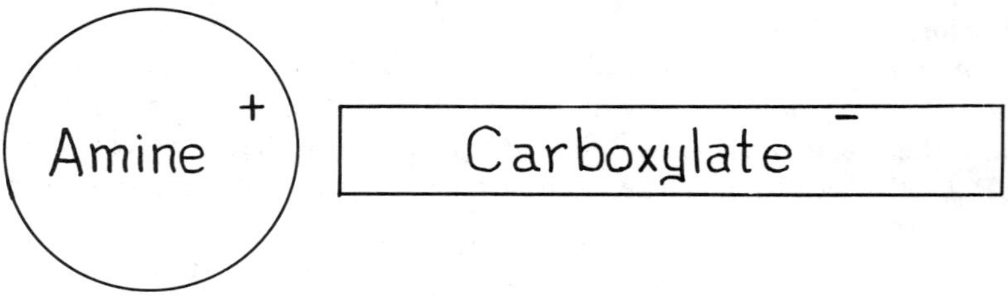

Fig. 3:5. Scheme of Amphoteric Surfactant

1. Amino and carboxy compounds.
2. Amino and sulfuric esters.
3. Amino and alkane sulfonic acids.
4. Amino and aromatic surfonic acids.
5. Miscellaneous combinations of acidic and basic groups.

Danielli[17] listed three other types of amphoteric surfactants:

1. Substituted betaines, e.g., N-dodecyl *n*:*n*-dimethyl betaine
2. Quaternary phosphono-ammonium compounds
3. Alkali salts of normal long-chain alkyl β amino propionic acid

Naturally Occurring Surfactants

The naturally occurring emulsifying agents were the first materials to be used and they still find wide application. Many are inefficient by themselves when compared to synthetic surfactants, but they are often useful as auxiliary emulsifying agents, dispersing agents, or protective colloids. Many of these surfactants materially increase the viscosity of an emulsion. Naturally occurring surfactants are generally somewhat expensive, they are subject to hydrolysis, and their efficiency is sensitive to variations in the *p*H of the system.[18]

Naturally occurring surfactants may be anionic, cationic, or nonionic. They include the alginates, steroids, lipids, water-soluble gums, and cellulose derivatives.

Alginates are the salts or esters of alginic acid and are derived from kelp.

Phospholipids and *steroids* are major constituents of many natural fats and oils.

Lecithin, a common phospholipid, is derived from soybean oil and is used as a stabilizer for oil-in-water emulsions.

Sterols, such as cholesterol, are derived from lanolin and used as stabilizers for water-in-oil emulsions.

Water-soluble gums, such as gum arabic, gum tragacanth, gum karaya, and guar gum, are principally complex polysaccharides. These gums are used as auxiliary emulsifiers and as such they both increase the fluid viscosity and form strong interfacial films.

Cellulose derivatives, e.g., methylcellulose and carboxymethyl cellulose also increase the viscosity of the emulsion.[19]

Oil-soluble Surfactants

The so-called *oil-soluble* surfactants are really composed of two major groups: Organo-silicone and fluorocarbon.

Organo-silicone surfactants are characterized by a silicone lipophilic group. These materials have found widest use as surfactants and additives for control of foam in the manufacture of polyurethane foams.

Anionic, cationic, and nonionic *fluorocarbon surfactants* have been developed. These relatively expensive materials often exhibit a synergism with conventional surfactants at low concentrations. The greatly lowered surface and interfacial tensions caused by these materials have promoted their use as leveling agents also. Whereas the surface tension of hydrocarbon surfactants may be as low as 30 dynes/cm, that of the silicones is in the low 20's; and that of fluorocarbons, in the high teens.

EMULSIFYING AGENTS

Emulsifying agents are primarily used to make emulsions. As redundant as that statement appears, it should be noted that the term "emulsifying agent" is broad.

Although nearly all emulsifying agents are considered surfactants, the converse is not true. Because of variation in cost, availability, and efficiency, not all surfactants are emulsifying agents.[20] But other surfactant or surfactant-like materials which may technically be dispersing agents, wetting agents, etc., are often classified in the trade as "emulsifiers" or "auxiliary emulsifier." The maxim in much of the trade appears to be "when in doubt, call it an emulsifier." A reference to an "emulsifier" may mean the surface-active agent, the mechanical equipment, or the person doing the emulsification. Throughout this book, the terms *surfactant* and *emulsifying agent* will be used instead of "emulsifier." Detergents are a specialized type of emulsifying agent designed to suspend or emulsify solids or oil in an aqueous media.

PENETRANTS AND WETTING AGENTS

Wetting is the process in which one fluid phase is displaced, completely or in part, from the surface of a solid or liquid by another fluid phase.[21]

Penetrants and wetting agents are surfactants that reduce the surface tension of water or reduce the interfacial tension between water and another surface. The interfacial tension may be between air and liquid, liquid and solid, or two liquids. These surfactants are usually quite effective in small amounts and will reduce the surface tension of pure water from 72 to about 30 dynes/cm. Wetting agents generally contain both polar and non-polar groups. The polar group is generally hydrophilic, whereas the non-polar portion of the molecule is lypophilic. The balance between polar and non-polar groups and their molecular position determine the characteristics of the wetting agent.

Penetrants and wetting agents are widely used in the textile industry to aid dyeing and sizing operations. Wetting agents are commonly evaluated by measuring the time required for a 5-gram skein of unboiled 2-ply cotton yarn to sink in a water dispersion of the wetting agent (Draves-Clarkson).[22]

The efficacy of a wetting agent is not universal. An agent may be effective in the presence of a lipophilic material, but small concentrations of a hydrophilic material may render the wetting agent inoperative. Therefore, the recommendation for a wetting agent or penetrant is conditional upon the exact nature of the system in which it is to be used.

Octyl alcohol, for example, can be used as a wetting agent for the following:[23]

Material to Be Wet	*Area of Use*	*% Alcohol by Wt.**
Pigment	Paint, Oil- or water-base	0.02-1.0
Pigment	Cosmetic lotions, water-based	0.001-0.01
Pigment	Lacquers	0.1-1.0
Pigment	Leather dressings, water-based	0.002-0.02
Casting Clays	Pottery	0.0001-0.001

* Based upon material to be wet.

Some of the more commonly used wetting agents and dispersants include the soluble salts of sulfated higher fatty alcohols, sulfonated amides, fatty acids, ester sulfonates, sulfonated ethers, alkylaryl sulfonates, and polyhydroxy alcohols partially esterified with higher fatty acids. Many emulsifying agents are also used as wetting agents. The materials in Table 3:1 are examples of the diversity of materials that are sold specifically as wetting agents.

TABLE 3:1

COMPOSITION OF WETTING AGENTS

Name	*Chemical Nature*	*Class**	*Supplier*
Aquarex SMD	Sodium methyl oleate sulfate	A	1
Aquarex D, MDL, ME, WAQ	Sodium alkyl sulfates	A	1
Arizona Surface Agent 351	Sulfated tall oil fatty acid	A	2
Nopco 2272-R	Sulfated fatty ester	A	3
Nopco 2031	Sulfated fatty acid	A	3
Aquarex L	Ammonium alkyl phosphate	A	1
Naccosol A	Sodium alkyl napthalene sulfonate	A	4
Nopco 9092	Alkylolamide plus alkyl aryl sulfonate	A	3
Hymic PE 90	Octyl phenol polyether alcohol	N	3
Hyonic GL-400	Ethoxylated octyl phenol	N	3
Nopco 1525	Alkylaryl polyether alcohol	N	3
Pluronic	Oxyethylene polyoxypropylene propylene glycol	N	5
Syntergent K	Modified coconut fatty acid alkanolamide	N	3
Alkaterge	Amine-type cationics	C	6
Modicol N	Fatty amine	C	3
Aquarex NS	C-cetyl betaine	X	1

* A—Anionic, N—Nonionic, C—Cationic, X—Amphoteric

Suppliers:

1. E. I. duPont de Nemours & Co.
2. Arizona Chemical Co.
3. Nopco Chemical Co.
4. National Aniline Div., Allied Chemical Corp.
5. Wyandotte Chemicals Corp.
6. Commercial Solvents Corp.

DISPERSING AGENTS

Through reduction of the cohesive attraction between particles, the dispersing agent will separate aggregations of particles. As a dispersing agent will not affect the dispersant medium, its surface tension and wetting effects may be minimal.

Protective colloids or wetting agents should not be used in conjunction with a dispersing agent because they will interfere with each other. Dispersing agents may be either organic or inorganic. Typical agents can be found in Table 3:2.

TABLE 3:2

COMPOSITION OF DISPERSING AGENTS

Name	*Chemical Nature*	*Supplier*
Daxad 23, 27	Sodium polymerized benzoid alkyl sulfonates	1
Daxad 21	Mono- and di-calcium polymerized alkyl aryl sulfonates	1
Daxad 11, 15	Sodium or potassium polymerized alkyl naphthalene sulfonates	1
Tamol	Sodium high-molecular-weight sulfonates or carboxylates	2
Daxad 30	Sodium polymerized carboxylate	1
Vinol PA-5, 523, 540	Partially acetylated polyvinyl alcohol	3
Naccotan A	Sodium naphthalene formaldehyde sulfonate	4
Solar S-2552	Modified alkanolamide	5
Marasperse	Lignosulfonates	6
Maratan	Sodium lignosulfonates	6
Norlig	Unmodified or partially modified lignosulfonates	6
Maracell	Lignosulfonates	6
Maracon	Lignosulfonates	6
Gypseperse	Lignosulfonates	6
Maracarb	Mixtures of lignosulfonic acid salts and salts of alkaline reversion of hexoses and pentoses	6
P&G Lignosulfonate	Ammonium lignosulfonate	7

Suppliers:
1. Dewey and Almy Chemical Div., W. R. Grace & Co.
2. Rohm and Haas Co.
3. Airco Chemical Div., Cumberland Chemical Corp.
4. National Aniline Div., Allied Chemical Co.
5. Swift & Co.
6. Marathon Div., American Can Co.
7. Procter & Gamble Co.

FOAMING AGENTS

A surface-active material that is used specifically to disperse gas in a liquid or solid medium is called a *foaming agent*, *foam booster*, *foam additive*, or *frothing agent*. In the production of a froth, the surfactant will tend to migrate to the interface and it may provide mechanical stability for the foam.

Foaming agents are widely used in consumer products as well as in industry. Shampoos and hand-dishwashing liquids must provide copious amounts of foam for aesthetic appeal. Frothing agents are used industrially in flotation processes to remove impurities from uranium and other ores.

PROTECTIVE COLLOIDS

A protective colloid is a material that stabilizes a particular dispersed system against aggregation, flocculation, or even coalescence, by such special means as added electrolytes or dehydrating agents, violent agitation, heat, or freezing. These colloids will usually form interfacial films of great stability, of a high degree of hydration, and frequently of great mechanical strength.[24] Therefore they tend to stabilize foams.

Polyvinyl alcohol, sulfonated naphthalene derivatives, alginates, alpha- and gamma-protein, casein, methyl cellulose, carboxymethyl cellulose, gelatin and glues, natural gums, polyacids, and starch are used as protective colloids.

DEFOAMING AND ANTIFOAMING AGENTS

As mentioned in Chapter 1, it is often advantageous or necessary to prevent the formation of foam or remove foam once it is formed. Anti- or de-foaming agents apparently replace the foaming agent at the interface. Therefore, the foam-eliminator must have intrinsically low surface tension, low solubility in the system, and good spreading characteristics.[25] Agents must spread rapidly and easily across the surface of the liquid or foam in order to be physically available for the task of controlling foam. A number of mechanical defoaming apparatuses are available which disrupt the foam through acoustical vibration. Chemical antifoaming or defoaming agents include:[26]

- Alcohols
- Aliphatic acids and esters
- Fatty acid soaps
- Halogenated compounds
- Natural fats, oil, and waxes
- Nitrogen-containing compounds
- Phosphates
- Silicone compounds
- Sulfate and sulfonates
- Sulfides and thio-compounds

The similarity of this list to the list of foaming agents is apparent. Many of the same materials will act as profoamers or defoamers depending on the system in which they are used. Because of this, defoamers and antifoamers are specific for certain systems. Many proprietary and compounded products are available. The amount of foam-controller that is necessary to for the job is a function of the foaming system, the de- or anti-foamer, and the amount of control desired. The foam in a boiling sugar solution is much harder to

TABLE 3:3

DEFOAMING AND ANTIFOAMING AGENTS

Name	*Chemical Nature*	*Class**	*Primary Use*	*Supplier*
Sul-fon-ate OA5	sodium sulfonated oleic acid	A	acid media	1
De-airex 535, 538	—	A		2
Arizona Surface Active Agent 351	sulfated tall oil fatty acid	A	paper	3
Antifoaming Agent D	high-molecular-weight fatty alcohols	N	paint	4
De-airex 506, 510, 516, 531, 540	blend of esters/hydrocarbons	N		2
De-airex 6041-F	esters/alcohols	N	paint	2
Foamtrol 103, 160	esters/hydrocarbons	N	broad range	5
45 Oil, Carnation, Pentol, Kaydol, Blandol, Benol, Protol	white mineral oils			6
Pluronics L61, 81, 101	oxyethylene polyoxypropylene propylene glycol	N	cleaners	7
Triton X15, 35, 207	alkyl aryl ether alcohol	N		8
Triton CF-32	amine polyglycol condensate	N		8
Atco-Airex	—			9
Arizona Surface Active Agent 301, 302	proprietary	A		3
Defoamer B Modified	—		yeast	10
Vegefat Y	—		beet sugar	10
Nopco series	—		broad range	10
Defoamers No. 1-1500	organic series of products		broad range	11
Defoamer No. 142	silicone	S	paper	11
Seyco Defoamer A69	silicone emulsion	S	textile	12
Brosco Antifoam LB	silicone	S	textile	13
Si-O-Kwel	silicone emulsions and compounds	S	broad range	14
Dow Corning Antifoam A,Q, 200 fluid	silicone compounds and fluids	S	broad range incl. foods	15
Dow Corning Antifoam AF, C, FG-10, A	silicone emulsions	S	broad range incl. foods	15

A—Anionic, N—Nonionic, S—Silicone

Suppliers:
1. Tennessee Corp.
2. E. F. Houghton & Co.
3. Arizona Chemical Co.
4. Henkel International GmbH (A. H. Carnes Co., agent)
5. Arlen Chemical Corp.
6. Utility Chemical Co.
7. Wyandotte Chemicals Corp.
8. Rohm & Haas Co.
9. Metro-Atlantic, Inc.
10. Nopco Chemical Co.
11. Swift & Co.
12. Seydel-Woolley & Co.
13. Scholler Brothers, Inc.
14. Stauffer Chemical Co.
15. Dow Corning Corp.

control than that in a cold dilute solution of one surfactant. Whereas 1 to 2% of a $0.10/lb organic defoamer may be required to defoam a system, 0.001 to 0.002% of a $2.00/lb silicone may yield comparable results. Foam-controlling agents are available as liquids and as grease-like compounds. Emulsions of foam-controllers agents are designed to invert or break when added to the foaming system. See Table 3:3 for a list of some of the available products.

A foam-controlling agent can be added to the foaming liquid or dispersed on one of the ingredients. Grease-like antifoaming compounds can be smeared on the sides or top of reactors or on a screen above the foaming liquid. The foam will be broken as it reaches the compound. Nozzles are coated with a grease-like antifoam to control foam during shampoo-filling operations without causing defoaming of the shampoo itself.

PRESERVATIVES

Emulsions have the capacity to support a flourishing population of micro-organisms. The combination of long-chain organic emulsifiers or additives and a moist medium provides an ideal breeding ground for both aerobic and anaerobic bacteria. The growth of micro-organisms ("bugs") will readily render the emulsion aesthetically unappealing. If the emulsifying agent happens to be the main course of the bugs' dinner, the emulsion will soon degenerate into a two-phase, rancid mixture.

The emulsion must be protected against three forms of degradation: growth of mould, growth of bacteria, and oxidation. The following conditions are necessary for the growth of moulds:

1. Access of air
2. Source of nutrient carbon and nitrogen
3. A moist environment

Whereas moulds can flourish over a wide range of *p*H, bacteria are destroyed by either acids or bases, but do not need air to survive. The fats and oils in the emulsion may oxidize into a rancid product.

Mateials that control growth of fungus or bacteria are called *fungistats* or *bacteriostats*. Those that destroy are called *fungicides* or *bactericides*. Common examples of such agents include: Benzoic acid, salicyclic acid, sorbic acid, benzoates, para-hydroxy-benzoates, phenylmercuric salts, phenols (including parachlorometacresol, parachlorometaxylenol, and dichlorometaxylenol), quaternary ammonium compounds, hexachlorophene, and dichlorophene.

Stuckey[27] compiled a list of modern, effective antioxidants from the permitted list of U.S. Foodstuffs.

PHENOLIC
- Butylated hydroxyanisole (BHA)
- Butylated hydroxytoluene (BHT)
- Guaiacum resin
- Lecithin
- Nor-dihydroguaiaretic acid (NDGA)
- Propyl gallate
- Tocopherols
- Trihydroxybutyrophenone

NON-PHENOLIC
- Ascorbic acid
- Ascorbyl palmitate
- Dilauryl thiodipropionate
- Distearyl thiodipropionate
- Monoisopropyl citrate
- Thiodipropionic acid

Emulsions that are to be used for foods, cosmetics, or drugs pose additional problems in preservation. These products must meet higher standards of preservation, but because of toxicity the most effective preservatives are often not allowed in such preparations.

Nonionic emulsions present still other problems. Because nonionic systems are generally neutral, bacterial growth can become a problem. In addition, many nonionic emulsifying agents reduce the effectiveness of preservatives such as the phenols and quaternary ammonium compounds. Although screening procedures and accelerated tests are available for determining the type and concentration of a preservative, aging the product formulation in the ultimate package is a universal procedure. The addition of the proper preservative will not substitute for sanitary procedures in the selection and use of the components of the emulsion. Generally, 0.1 to 0.2% of preservatives such as sorbic acid, benzoic acid, and phenols are added routinely to emulsions for minimal bacteriostatic protection.

Following is a partial list of the suppliers of antioxidants, bactericides and fungicides:

Advance Division, Carlisle Chemical Works

The Dow Chemical Co.

E. I. duPont de Nemours & Co. Inc. (Polychemicals Dept.)

Eastman Chemical Products, Inc.
Geigy Industrial Chemicals
Heyden Division, Tenneco Chemicals, Inc.
Hooker Chemical Corp.
Monsanto Co.
M&T Chemicals, Inc.
Chas. Pfizer & Co., Inc.
Reheis Chemical Division, Armour Pharmaceutical Co.
Rohm and Haas Co.
R. T. Vanderbilt Co.
Sindar Division, Givaudan Corp.

References

1. MOLLIET, J. L., COLLIE, B., and BLACK, W., *Surface Activity*, (2nd Ed), pp. 5-6; D. VAN NOSTRAND CO. Inc., PRINCETON, N.J., 1961.
2. *Ibid.*, pp. 3-5.
3. SCHWARTZ, A. M., and PERRY, J. W., *Surface Active Agents*, pp. 18-20; INTERSCIENCE PUBLISHERS, Inc., NEW YORK, 1949.
4. BECHER, P., *Emulsions, Theory and Practice*, pp. 216-8; REINHOLD PUBLISHING CORP., NEW YORK, 1965.
5. Spalton, L. M., *Pharmaceutical Emulsions*, pp. 11-14; CHEMICAL PUBLISHING CO., INC., NEW YORK, 1950.
6. BECHER, *op. cit.* (4) pp. 218-220.
7. *Ibid.*, pp. 213-5.
8. *Ibid.*, pp. 216-8.
9. Cationic Surfactants, *Drug and Cosmetic Ind.*, 92:368, (1963).
10. *Ibid.*
11. BECHER, *op. cit.* (4), p. 221.
12. *Ibid.*, p. 221.
13. *Ibid.*, pp. 222-3.
14. *Ibid.*, pp. 223-5.
15. *Ibid.*, p. 225.
16. *Ibid.*, pp. 209-211.
17. DANIELLI, J. F., PANKHURST, K. G. A., RIDDIFORD, A. C., Editor, *Recent Progress in Surface Science*, p. 266; Academic Press, Inc., (1964).
18. BECHER, *op. cit.* (4), pp. 209-211.
19. *Ibid.*, p. 226.
20. *Ibid.*, p. 212.
21. MOLLIET, *op. cit.* (1), p. 132.
22. DRAVES, *Am. Dyestuff Reptr.*, 28:425 (1939).

23. BENNETT, *Chemical Formulary*, IV: 100: CHEMICAL PUBLISHING CO., INC., NEW YORK, 1939.
24. MOLLIET, *op. cit.* (1), pp. 177-8.
25. MOLLIET, *op. cit.* (1), pp. 206-7.
26. BECHER, *op. cit.* (1), p. 147.
27. STUCKEY, B. N., *American Perfumer*, 73 (2):35 (1959).

chapter 4

ANALYSIS AND TESTING OF EMULSIONS

A wide variety of tests is available for emulsions. The formulator uses these tests to characterize and analyze his products and those of his competition. Many of these tests are used for quality control to ensure that the emulsion at hand corresponds to the emulsion that was produced in the laboratory. Some tests are applicable for the determination of projected shelf stability of the emulsion. Other tests are specifically designed for academic characterization of the emulsion or ascertaining the effect of a change in ingredients upon the properties of the emulsion.

TESTS FOR TYPE OF EMULSION

Electrical Conductivity

The electrical conductivity of an emulsion depends on that of the continuous phase. Water is a good conductor, and most oils are poor conductors. W/O emulsions in which the dispersed phase amounts to 60% by volume or more have moderately high conductivity.

Ehrlich's Filter Paper

When an emulsion is brushed on a piece of filter paper, a wet margin may appear. If it is a wide margin, an O/W emulsion is present; if it is narrow, a W/O emulsion. A mixed emulsion is difficult to recognize.

Direction of Creaming (Becher)

If the relative densities of the two phases are known, the direction in which creaming of the dispersed phase takes place will indicate the type of emulsion present. Normally, creaming is upward for O/W emulsions, and downward for W/O emulsions. This method assumes that the oils involved are less dense than water; otherwise the converse would be true.

Dye Solubility

This method employs dyes having specific solubilities and is based on the fact that a dye will tint only the continuous phase uniformly. Oil-soluble dyes are absorbed by W/O emulsions only. Of two small samples, one is sprinkled with some grains of an oil-soluble dye, and the other with some grains of a water-soluble dye. The penetration of the dyes can be observed from the formation of a colored ring in one sample; formation of a ring in both samples indicates a mixed emulsion.

Another procedure involves pouring an aqueous solution of the testing dye over one sample of the emulsion and a dye dissolved in oil over another sample. After some minutes, the dye solutions are discarded and the samples washed briefly with their respective solvents. The color of the emulsion will reveal which dye had penetrated.

Water-dispersibility

This is suitable for low-viscosity emulsions. O/W emulsions can be stirred with water, but will not mix with oil; the reverse holds for W/O emulsions. Mixed emulsions can be stirred with both oil and water to some extent. The oil that is used must be non-emulsifying.

Fluorescence

Many oils fluoresce under ultraviolet light. A drop of emulsion is examined under a microscope in fluorescent light. If it is a W/O emulsion the entire field will fluoresce; in an O/W emulsion fluorescence is spotty.

Light scattering

A parallel bundle of light rays, going through a droplet of O/W emulsion, will be converged; through a W/O emulsion, it will be diverged.

TESTS FOR DENSITY OF EMULSION

The density (or specific gravity) of low-viscosity liquids is determined by a *hydrometer*. A *pycnometer* is used for materials of higher viscosity, For slurries, the following relations apply:

$$P_v = \frac{D_m - D_l}{D_s - D_l}$$

$$D_m = \frac{1 - P_v}{1 - P_w}$$

where P_v = fraction by volume of solids
P_w = fraction by weight of solids

$$= \frac{\text{weight of solids}}{\text{weight of mixture}}$$

D_l = density of liquid

D_m = density of mixture

D_s = density of solids

TESTS FOR VISCOSITY OF EMULSION

The determination of viscosity is one of the most common tests in quality control and for the general characterizing of an emulsion. Viscometers measure either the time for a fixed volume to flow through an orifice, or the force or *torque* required to maintain a fixed rate of shear. All measurements of viscosity should be made at a constant, standard temperature (e.g., 20 or 25°C) in order to permit the comparison of results. The method of determining viscosity should also be specified. A comparison of the torques resulting from two different rates of shear can be compared to yield an estimate of the type and amount of non-Newtonian fluid characteristics.

The five types of viscometers include:

1. Poiseuille type
   - Capillary viscometers
   - Orifice viscometers
2. Stokes type
   - Falling ball viscomeers
   - Rising bubble
   - Piston
3. Couette type
   - Paddles
   - Concentric cylinders
4. Newton type
   - Sliding plate
5. Coulomb type
   - Damping rate

Poiseuille–Type Viscometers

The original capillary viscometer developed by Ostwald for measuring the kinematic viscosities of Newtonian liquids has undergone a number of modifications. The units suitable for transparent liquids of low viscosities include *Ubbelohde*, *Cannon-Fenske*, *Fitzsimmons*, *Zeitfuchs*, *SIL*, and *Atlantic*. Modifications have resulted in the following designs, which are suitable for opaque liquids: Cannon-Fenske Reverse Flow, Zeitfuchs Crossarm, and Lantz-Zeitfuchs.

ASTM D445-65, *Kinematic viscosity*, gives a general description of methods and designs for these units. All are based on observation of the efflux time under specified conditions. *Apex*, *Cox*, and *Severs* capillary viscometers are used for high-consistency, low-flow fluid systems. The sample is extruded under pressures up to 2000 *psi* in order to determine the absolute viscosity.

The typical orifice viscometer consists of a cup with an orifice of known diameter in the bottom. The cup is filled with the material to be tested. The time required to empty the cup is recorded and converted into kinematic viscosity by a calibration chart.

The *Engler* viscometer (ASTM D490) has a bronze inner cup and a platinum outlet tube. The Engler viscometer is mounted in a water bath for temperature control.

The *Ford* viscosity cup, a modification of the Engler unit, is described in ASTM D1200.

The *Saybolt* orifice viscometers come in three sizes to suit a wide variety of materials. The *Saybolt Universal* viscometer (ASTM D88-53) is used for oils of moderate viscosity. ASTM D446-53 permits ready conversion of Saybolt Universal seconds into kinematic viscosities (centistokes) through the table provided in that specification. A Saybolt viscometer fitted with a Furol tip reduces the flow times by an order of magnitude. Table 1 in ASTM 666-53 gives the kinematic viscosities between 48 and 400 centistokes as a function of the Saybolt Furol seconds at 50°C. The efflux time of Asphalt tip, in turn, is about 1/10 that of the Furol tip.

The Zahn viscometer is a bullet-shaped metal cup of 45 cm^3 capacity, with a calibrated orifice. Zahn viscometers numbered 1 to 5 are used for viscosities of 20 to 2000 centipoise.

Stokes–Type Viscometers

The *Hoeppler* falling-ball viscometer determines absolute viscosity from the time required for a steel ball to fall a specified distance through the fluid. A water-jacket maintains a constant operating temperature of 20°C. Fluids with viscosities up to one million centipoise can be characterized with a Hoeppler viscometer.

The *Gardner Timer Tube* which is used for transparent liquids, is described in ASTM D1545-63. The tube has a flat bottom at one end and is open at the other. It has two marked rings 73 mm apart. The time for a bubble to travel through this distance upon inversion of the tube is a measure of the viscosity of the fluid. The "Gardner seconds" are roughly comparable to stokes.

In the *Norcross* viscometer, a piston is raised within a cylinder and is allowed to fall by gravity. The time of fall of the piston a measure of the viscosity of the liquid. The *Mobilometer* is similar in principle to the Norcross viscometer and is used to determine the viscosity of liquids up to 500,000 centistokes.

Couette Viscometers

The *Stormer* concentric-cylinder viscometer measures the time required for a given number of revolutions of a cylinder immersed in the sample of fluid. The relative and absolute viscosities are then found from a calibration table. In determining the consistency of paints or other thin liquids, the rotating cylinder can be substituted by a paddle-type rotor (ASTM D562).

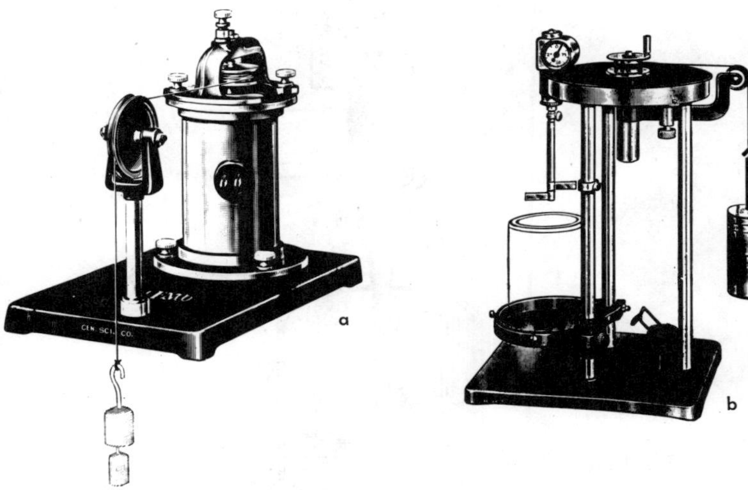

Fig. 4:1. TYPES OF VISCOMETER
a) Rotating Cylinder (*Cenco*, *Inc.*); *b*) Stormer Model.

Brookfield viscometers measure the resistance to rotation at a fixed *rpm* in the liquid. Different spindle-and-speed combinations provide for the measurement of the viscosity of a wide range of liquids.

Newton-Type Viscometers

In the *Hallikainen-Shell Sliding Plate* viscometers, the sample is placed between two glass plates. The shear force is determined by pulling one plate away from the other at a fixed rate. The absolute viscosity, a function of the pulling force, can thus be determined for materials with viscosities in the range of 10^3 to 10^{11} centistokes. Both Newtonian and non-Newtonian fluids can be evaluated on this device.

Coulomb-Type (Damping Rate) Viscometers

In the *Coulomb-Type* device, a metal reed, clamped at one end and free to vibrate at the other, is vibrated at its natural frequency by an electromagnet. As the vibrating reed is immersed in a liquid or semisolid, its vibrating frequency is reduced. The resulting variation in the exciting voltage is a measure of the viscosity of the liquid. This principle is applied in the *Agfa* meter. The oscillating viscosity tester developed by the Bell Telephone Laboratories is another Coulomb type of viscometer.

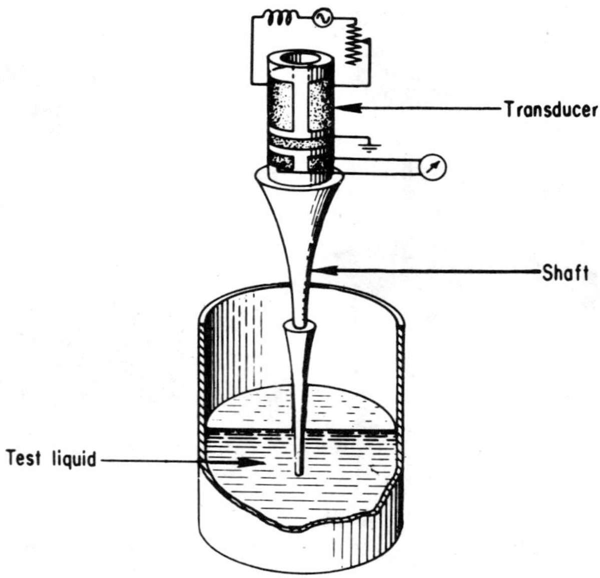

Fig. 4:2. Oscillating Viscosity Tester
(U.S. Patent 3,131,515; *Bell Telephone Laboratories*, New York)

TESTS FOR SURFACE TENSION

Seven methods for determining surface tension were outlined by Osipow.[1] Measurements of surface tension are often useful in the evaluating of surfactants and surfactant blends. The *capillary rise* and the *ring* methods are the most commonly used.

Capillary Rise Method

The classical technique for determination of surface tension is the *capillary rise* method.

One end of a capillary tube is immersed in the liquid to be tested. The liquid in the capillary will rise or fall somewhat above, or below, the level of the surrounding liquid. The surface tension is a function of the angle of contact, and directly proportional to the radius of the capillary, the difference in the change in pressure, the difference in density between the gaseous and the liquid phases, and the amount of rise or depression of the liquid in the capillary.

Ring Method

The *ring method* is the most widely used commercial method for the rapid and accurate determination of interfacial or surface tension. It measures the force required to detach a thin ring from the surface or interface of the liquid. The surface or interfacial tension is inversely proportional to the radius of the ring, and directly proportional to the sum of the total indicated weight, the weight of ring, and a correlation factor.

Wilhelmy Slide Method

A thin platinum plate is suspended from one arm of a balance, dipped into the liquid, and gradually raised to the point of detachment from the interface. The surface or interfacial tension is a function of the total indicated weight, the weight of the slide, and inversely proportional to the width and thickness of the slide. The slide must be wet by the liquid to use this method.

Maximum Bubble Pressure Method

The maximum pressure required to force a bubble from the short arm of a U-tube under the surface of the liquid is measured. The surface tension is a function of the radius of the tube and of the difference in pressure between the gas that forms the bubble and the liquid.

Drop-Weight Method

Individual drops of the liquid are allowed to fall from the end of a vertical tube. Both surface and interfacial tension are inversely proportional to the radius of the tube, and directly proportional to the weight of the drop and a correction factor.

Static Drop Method

The change in shape of a drop of liquid is a function of slow changes in the surface tension. This method is often used for molten metals and in other cases when the changes in surface tension are of interest.

The *Sessile Drop* is a modification of this method for industrial use. The surface tension of the liquid is a function of the square of the height of a drop placed on a flat plate.

Jet Method

The *Dynamic Jet* method allows the study of rapid changes in either the surface tension or the interfacial tension of a liquid. After a liquid is extruded from a non-circular orifice, the surface or interfacial tension causes the drop to attempt to form a circular configuration. The momentum of this action, however, causes overshooting and results in subsequent cyclical changes in the shape of the jet of liquid.

The surface or interfacial tension is inversely proportional to the radius of the circular cross-section of the jet, and to the square of the wave length of oscillation. It is directly proportional to the density of the fluid and to the square of the rate of flow.

TESTS FOR PARTICLE SIZE

The average size of the particles in an emulsion and the particle size distribution can be used to determine the efficacy of the mixing procedure. "Guestimates" of the shelf stability of an emulsion are also possible through knowledge of particle sizes.

Orr and Dallavalle list nine types of equipment for determination of particle size.[2]

Determined by	*Particle Radius*
Optical Microscope	0.25—250μ
Electron Microscope	0.0005—5μ
Liquid Sedimentation	0.5—300μ
Gas Sedimentation (Eluriation)	1.0—200μ
Hindered Settling	5.0—1000μ
Centrifugal Sedimentation	0.0005—50μ
Impactation	0.1—100μ
Light Scattering and Transmission	0.1—30μ
X-ray Scattering	0.005—0.05μ

In addition, biological colony counters are now being used for determining particle size. The Coulter Counter is an example of a semi-automatic device which is used for determining particle sizes in emulsions.

Microscopes

The use of microscopes to determine particle size requires a thin film of the emulsion and often results in agglomeration or coalescence during the preparation of the sample.

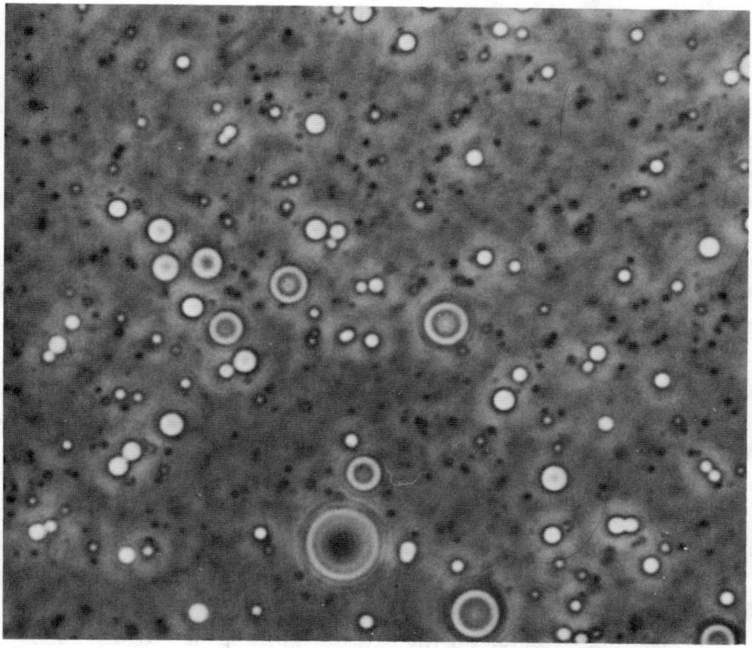

Fig. 4:3. PHOTOMICROGRAPH OF AN OIL-IN-WATER EMULSION
This is a hand lotion, magnified 1100 times, by phase contrast technique. Blurring of small particles due to Brownian Movement. (From *Dow Corning Corporation*, Midland, Michigan)

The *optical microscope* is one of the more commonly used pieces of equipment for of particle size determinations. Coalescence is reduced by adding a concentrated solution of gelatin. The sizes of globules are determined with a squared eyepiece calibrated by means of a stage micrometer. Groups of various sizes are counted and compared with counts of similar samples of emulsion. To minimize errors, at least 1000 globules should be counted, using at least two slides and several noncontiguous fields on each slide. The focus is varied to examine different depths of the film.

With an *electron microscope*, specimens must be examined in a vacuum; therefore they must be dry. The sample is transferred to a glass slide and scraped off with a steel strip, which leaves ridges of emulsion. The oil phase is leached out of the emulsion with a suitable solvent, and by vacuum evaporation the specimen is coated with a layer of suitable material. The shadow effect so obtained improves contrast, but the internal details of the particle are lost. Particle count may also be from photomicrographs, which is more convenient. Montgomery pointed out that light-scattering, turbidity, soap titration, electronic particle-counters, creaming rates, and ultracentrifuges do not distinguish between single particles and agglomerates. He described the use of automatic and semiautomatic particle counters in conjunction with the electron microscope. The semiautomatic counter reduces the tedious business of counting a large number of particles.[3]

Diffraction or Scattering

The presence and quantity of additives in emulsions can be determined from the arrangement of atoms within solid particles and their interrelation. The diffraction method (X-ray or electron) gives no direct information on sizes of particles, but it does indicate chemical identities and the distribution of atoms in the material.

The size of the particles can be appraised from the scattering obtained. *Mie scattering* occurs with particles of diameters between 0.33 and 20 times the wave length of the radiation. For particles with diameters less then 0.33 λ, *Rayleigh scattering* occurs.

Mie scattering depends on the ratio of particle size to wave length, and is proportional to the density of particle-scattering. Rayleigh scattering is a function of λ^{-4} and of the density of particle-scattering. Data on diffraction can be correlated with electron micrographs to give useful results.

Settling and Sedimentation

The various methods of sedimentation or settling are based upon Stokes Law. The settling (or rising) of the particles in an emulsion is a function of the size of the particle. The amount of sedimentation is proportional to the number of particles of any size. The individual methods use different procedures to increase the rate of sedimentation and determine the amount and size of the sediment.

Other Methods

Thacher[4] proposed a method of analyzing particle size based on the dielectric constant. He felt that the dependence of the dielectric increment on the strength of the electrical field might be used to investigate the distribution of size of particles in an emulsion. Some changes in the dielectric absorption, due to the distorting of the droplet by the electrical field, might be related to the size of the droplet.

TESTS FOR pH VALUE

The pH of a substance is the negative logarithm of the hydrogen ion concentration. pH may be measured by either colorimetric or electrometric methods. Many indicators are available that change color at specific concentrations of acid or base. A specific liquid or combination of liquids will indicate the approximate pH of the emulsion through change in color. Paper strips are also available, impregnated with a series of liquids that change the color of the paper at various pH levels. The change in color can be compared with tinted glass or paper standards that are labelled with the corresponding pH values. These colorimetric methods are occasionally affected by the presence of colloids but generally they give a rapid, inexpensive, reasonably accurate approximation of the pH of the emulsion.

Electrometric methods of determining pH are more expensive than the colorimetric methods, but much more accurate. The voltage between two electrodes in contact with the solution is a function of the pH of the emulsion. pH can be measured accurately, both continuously and occasionally, with electrometric pH meters. A single glass electrode is commonly used, and commercial instruments for this purpose are available from Corning Glass Works, Beckman Instruments, and Fisher Scientific Supply, among others.

TESTS FOR WATER CONTENT

The amount of water in an emulsion may be determined by Karl Fischer titration; this requires a titrating adapter to the pH meter. (See discussion in E203-62T, under *Water*, in section on *ASTM Specifications*.)

TESTS FOR COLOR

Where the color of a product is important, arbitrary scales such as the Saybolt, Tag-Robinson, and Gardner color scales are used. Numbers on each scale correspond to a series of colors on glass, plastic, or paper slides. Self-made standards are also common. Two jars or bottles of the emulsion indicating the widest allowable variation in the color of the product are used as a basis of comparison for the emulsion under consideration.

TESTS FOR ODOR

Tests are sometimes used to determine the cause of objectionable odors. Although Gas-Liquid or Vapor-Phase Chromatographs (GLC or VPC) are sometimes used for odor control, the human nose is usually the best analytical imstrument. For comparative purposes, standard odors should be maintained in the laboratory.

TESTS FOR CORROSION

Tests are often run to ascertain whether the emulsion will cause undesirable corrosion of containers. Small pieces of metals or plastics are placed in glass bottles containing the emulsion. Corrosion inhibitors are added to some of the emulsions to check their protection of the materials. The bottles are often stored at an elevated temperature, e.g., 50°C as well as at room temperature. Corrosion can be determined by visual observation or by loss of weight in the piece of metal or plastic.

TESTS FOR STABILITY

The stability of the emulsion should be evaluated in the package in which the product is to be sold, and under adverse as well as typical storage conditions. An emulsion should retain its appearance, viscosity, particle size, odor, and type over a reasonable period of time. The shelf life can be determined by aging or by accelerated tests, such as the use of a centrifuge. The samples can be centrifuged until separation occurs. To estimate the stability of the emulsion, the time at which separation occurs is compared to with that of a known standard.

Temperature

The stability of an emulsion can also be affected by temperature. The cooling and freezing, or heating and boiling, of an emulsion cause special problems in relation to stability.

Cooling an emulsion until the aqueous phase freezes will often cause separation. The film of emulsifier around the droplets in the dispersed phase may burst or be damaged by subsequent thawing. The solubility balance of the surfactant may also change with a decrease in temperature. Slow freezing is more damaging to the emulsion than rapid freezing as larger ice crystals are formed during slow freezing. The resistance of an emulsion to freezing is improved by the use of a water-soluble additive (e.g., ethylene glycol) that depresses the freezing point of the aqueous phase.

An emulsion is tested for stability to freezing by exposing it to a series of "freeze-thaw" cycles; in the paint industries, a common test consists of 16 hours at a temperature of —12°C followed by 8 hours at 24°C. The product is expected to withstand a certain number of such cycles before the emulsion coagulates. Samples of an emulsion may also be placed in a refrigerator and the temperature held near the freezing point for several days to determine the stability of the emulsion at near-freezing temperatures.

Heating lowers the viscosity of the external phase of an emulsion and may cause discontinuities in the film of the emulsifying agent. It also changes the solubility balance of the surfactant and facilitates chemical reactions.

The effect of heat on an emulsion may be evaluated in several ways: (1) samples can be held at a constant temperature of 45° to 50°C for a few days or months. Such samples are checked periodically by the quality control tests applied to freshly made emulsions. An increase of ten degrees centigrade in the temperature is considered to double the rate of most reactions. Therefore, three months at 45° to 50°C is equivalent to one year at 20° to 25°C for many systems.

(2) Another method of evaluating the effect of heat on the emulsion employs a glass cylinder. The glass cylinder is 25 mm in diameter with a bulbous extension at the bottom; it is fitted with a side cock, and rests in a wooden support. One hundred cc of the emulsion in the cylinder is placed in a drying oven to settle at a definite temperature and time. After 20 cc of the emulsion is removed and discarded, 10 cc are withdrawn into a tared aluminum beaker and weighed before and after evaporation in a drying oven. This gives the moisture content. If the emulsion contains B percent of the dispersed phase, a is the percentage of the dry substances of the phase, and b is the weight of the evaporated moisture; the content of the non-fatty part in the emulsion is:

$$\frac{b(100+a)}{a}$$

and the stability of the emulsion is:

$$\frac{b(100+a)}{B}$$

(3) The glow of an electric bulb is used to indicate the breaking

point of an emulsion in a third test. The emulsion is placed between two copper plates that are kept apart by two pieces of glass 0.2 mm thick. The bulb is attached to the copper plates and an alternating current is applied to the copper plates. The change in conductivity indicates the point at which the emulsion is destroyed. The breaking current can be compared with a set of standards to indicate the stability of the emulsion.

Effect of Dilution

To determine the smallest concentration of surfactant that will ensure stability, the emulsion is gradually thinned by the addition of more of the external phase. Dilution is continued until the emulsion breaks or creams. A specified stability to dilution may be required of agricultural emulsions as well as other products. The emulsion may also be diluted with a fixed amount of the continuous phase and a soluble dye. After centrifuging, the dye will indicate the degree of creaming or breaking in the emulsion. The amount of creaming can be related to the stability of the emulsion.

TESTS FOR CONTAMINATION

Contamination of an emulsion may arise from either the process of manufacture or the raw materials. The presence of such impurities in the emulsion may not be obvious during the manufacture because they may not react with the processing equipment or interfere with the process itself. Such contaminants, however, may interfere with the acceptability of the emulsion so it is often desirable to monitor the emulsion for such contamination.

Particulate impurities can usually be detected under the optical microscope. Dissolved or finely divided dispersed contaminants can be analytically found by the Gas-Liquid (Vapor-Phase) Chromatograph, infrared or ultraviolet spectrophotometer, etc.

TESTS FOR PERFORMANCE

Tests should be run on the final product and package to verify the efficacy of emulsion for its ultimate use. In general, such performance tests are designed to simulate the intended application of the product. With this in mind. excellent performance tests can be developed that are limited only by the originality of the inventor.

Some performance tests, as listed by the American Society for Testing and Materials (ASTM), are appended to this chapter. Often a customer will specify performance tests and test methods. For example, the Federal Government maintains a file of specifications which the products it purchases must meet or exceed.

ANALYTICAL PROCEDURES

Frequently the formulator of emulsions will wish to have a chemical analysis of the contents of an emulsion. Conventional analytical procedures are adequately covered in standard texts, but a few procedures specifically designed for the identification and characterization of surfactants are pertinent.

Rosen published an excellent review of the literature on the analysis of surfactants.[5] He listed three major chromatographic, qualitative analytical procedures. *Ion-exchange* resins can be effectively used to separate ionic from nonionic surfactants. *Paper chromatography* is used for the separation and/or identification of individual surfactants. In order to use a Gas-Liquid (Vapor-Phase) Chromatograph (GLC-VPC), the surfactant must be chemically converted to a more volatile derivative. The GLC can then be used to determine the structure of a surfactant and any impurities present, and establish the identity of the material and the supplier.

Desmond and Borden described the identification of surface-active agents in admixtures by thin-layer chromatography. Both qualitative and semi-quantitative identification of alkylaryl sulfonates, soaps, xylene sulfonates, toluene sulfonates, sulfated alcohol ethoxylates, sulfated alkylphenol ethoxylates, amine oxides, alkanolamides, and ethoxylates were described.[6]

Many analytical techniques first require the separation of the components of a surfactant blend. Blank and Kelly[7] described the following eight separations:

1. Soap from synthetic surfactants
2. Sulfonates of high and of low molecular weight
3. Alkylaryl sulfonates from fatty alcohol sulfates
4. Alkylaryl sulfonates from sulfated fatty acid monoglycerides
5. Alkylaryl sulfonates from nonionic surfactants
6. Fatty alcohol sulfates from sulfated fatty acid monoglycerides
7. Fatty alcohol sulfates, alkylaryl sulfonates, and alkanolamides
8. Alkylaryl sulfonates, polyoxyethylated anionic sulfates, and nonionics

Analytical procedures are also described in various publications of professional organizations, manufacturing companies, and standards associations, such as:

American Society for Testing and Materials
1916 Race Street
Philadelphia, Pa. 19103
Association of Official Agricultural Chemists, Inc.
Box 540, Benjamin Franklin Station
Washington, D.C. 20044
The Toilet Goods Association, Inc.
1270 Avenue of the Americas
New Yok, N.Y. 10020

ASTM SPECIFICATIONS PERTINENT TO THIS SUBJECT

The following specifications and testing methods have been published by the American Society for Testing and Materials. This list is only a sampling of ASTM Tests, which constantly undergo re-evaluation as new methods are developed.

SOAPS

D2179-65	Bar Soap, Combination
*D497-52	Bar Soap, Ordinary Laundry
*D496-51	Chip Soap
*D690-48	Chip Soap, Compound (with Rosin)
*D630-42	Chip Soap, Olive Oil (Type A, Straight; Type B, Blended)
D1111-53	Chip or Granular Soap for Low-Temperature Washing (Low and Medium Titer)
*D534-42	Powder Soap, Alkaline
*D533-44	Powdered Soap, Built
*D691-44	Powdered Soap, Compound (Granulated with Rosin)
*D498-51	Powdered Soap (Nonalkaline Soap Powder)
*D593-42	Salt-Water Soap (Coconut Oil Type)
D1112-53	Solid Soap for Low-Temperature Washing (Low and Medium Titer)
*D455-55	Toilet Soap, Milled
*D499-48	Toilet Soap, White Floating

* Specifications so marked have been approved as American Standard by the American Standards Association.

*D799-62	Toilet Soap, Liquid
	ALKALINE DETERGENTS
D929-50	Borax
*D458-57	Soda Ash
*D456-39	Soda, Caustic (Anhydrous)
*D457-39	Soda, Modified (Sesquicarbonate Type)
D928-52	Sodium Bicarbonate
*D537-57	Sodium Metasilicate
*D594-41	Sodium Sesquisilicate
*D538-60	Trisodium Phosphate
*D595-45	Tetrasodium Pyrophosphate (Anhydrous)
	SYNTHETIC DETERGENTS
D2021-65	Neutral Detergent, 40 Percent Alkylbenzene Type
D2356-65T	Scouring Powder Compounds
	VISCOSITY AND CONSISTENCY
D2162-64	Master viscometers and viscosity oil standards
D1745-63	Calculation of absolute viscosity
D2161-66	Conversion of kinematic viscosity to Saybolt Universal or Furol
STP 43-B	Viscosity tables for kinematic viscosity conversions and viscosity index calculations
D445-65	Viscosity of transparent and opaque liquids; (kinematic and dynamic)
D88-56	Saybolt viscosity
D341-43	Viscosity-temperature charts for liquid petroleum products. These charts give the viscosity of a petroleum oil at any temperature within a limited range, when viscosities at two other temperatures are known.
D567-53	Calculating the viscosity index. The viscosity index is an empirical number indicating the effect of a change of temperature on the viscosity of an oil. A low viscosity index signifies a relatively large change of viscosity with temperature.
D1298-55	Specific gravity of petroleum liquids, hydro-

	meter method
STP 42	Saybolt viscosity index tables
STP 168	Kinematic viscosity index tables
E201-62T	Calculation of volume and weight of industrial chemical liquids
D1092-62	Apparent viscosity of lubricating greases
D553-42	Viscosity and total solids content of rubber cements
D1243-66	Viscosity of vinyl chloride polymers
D1823-66	Apparent viscosity of plastisols and organosols at high shear rates by Castor-Severs viscometer
D1824-66	Apparent viscosity of plastisols and organols at low shear rates by Brookfield viscometer
E143-61	Determination of shear modulus at room temperature
D1343-56	Viscosity of cellulose derivatives by ball-drop method
D1795-62	Intrinsic viscosity of cellulose
D1725-62	Viscosity of resin solutions
D562-55	Consistency of paints, using the Stormer viscosimeter
D1200-58	Viscosity of paints, varnishes and lacquers by Ford viscosity cup
D1545-63	Viscosity of transparent liquids by bubble time method
D1311-61	Viscosity reduction power of hydrocarbon solvents
E102-62	Saybolt Furol viscosity of bituminous materials at high temperatures

PARTICLE SIZE

D1210-54	Fineness of dispersion of pigment-vehicle systems
D1316-54T	Fineness of grind of printing inks by grindometer
D869-48	Evaluating degree of settling of traffic paint
D1309-56	Settling properties of traffic paints during storage
E161-60T	Precision micromesh sieves

E20-62T	Particle size distribution of particulate substances of subsieve sizes: analysis by microscopical methods
D185-45	Coarse particles in pigments, pastes, and paints
D1366-65	Practice for reporting particle size characteristics of pigments
D502-39	Soaps and other detergents particle size

DENSITY AND SPECIFIC GRAVITY

D1895-65T	Apparent density, bulk factor, and pourability plastic materials
D1217-54	Density and specific gravity of liquids, Bingham pycnometer
D941-55	Density and specific gravity of liquids, Lipkin pycnometer
D1963-61	Specific gravities of drying oils, varnishes, resins, and related materials
D1481-62	Density and specific gravity of viscous materials by Lipkin pycnometer

COLOR

D307-44	Luminous reflectance, transmittance, and color of materials
D1209-62	Color of clear liquids (platinum-cobalt scale)
D1544-63T	Color of transparent liquids (Gardner color scale)

*p*H

E70-52T	Determination of *p*H of aqueous solutions with glass electrode
D1172-56	Soaps and Detergents: *p*H of aqueous solutions

REFRACTIVE INDEX

D1747-62	Refractive index of viscous materials

CHEMICAL ANALYSIS OF SOAPS AND DETERGENTS

*D501-58	Alkaline detergents: sampling and chemical analysis
D501-62T	Sodium triphosphate: sampling and analysis
D1568-63	Alkylbenzenesulfonates: sampling and chemi-

	cal analysis
D2024-65	Cloud point of nonionic surfactants
D1569-62	Detergent alkylate
D1767-62	Ethylenediaminetetraacetate (EDTA) in soap or synthetic detergents
D1570-63	Fatty alkyl sulfates: sampling and chemical analysis
D2357-65T	Qualitative classification of surfactants by infrared absorption
D2358-65T	Separation of active ingredient from surfactant and syndet compositions
D460-60	Soaps and soap products: sampling and chemical analysis
D820-58	Soaps containing synthetic detergents: chemical analysis

ANALYSIS OF WATER

E203-64 Test for water, using Karl Fischer reagent. This test is widely used for plastics, paints, petroleum products, drugs, food-stuffs, explosives, and naval stores. The percentage water is

$$P=100\ (FV/W)$$

where:

V=Volume of Karl Fischer reagent, ml
F=Concentration, mg water/ml reagent
W=Weight of sample, mg

*The method was published originally in *Angew. Chemie* 48:394. (1937). A nomograph is presented in *Chem. Processing*, August, 1965, pp. 88-90.

D1364-64 Water in lacquer solvents and diluents

MISCELLANEOUS RECOMMENDED PRACTICES

E177-61T	Use of the terms *precision* and *accuracy* as applied to the measurement of a property of a material
E122-58	Sample size to estimate the average quality of a lot or process

TESTING OF PRODUCTS

WAX POLISHES

D1791-66	Accelerated aging of liquid water-emulsion floor polishes
D1386-59	Acid number (empirical) of natural waxes
D1436-64	Application of emulsion floor polishes to substrates for testing purposes
D1288-57	Ash, total, and silica in water-emulsion waxes
D1437-64	Flash point of solvent-type liquid waxes
D1455-64	Gloss, 60-Deg. specular, of emulsion floor polish
D2046-64T	Nonvolatile matter (total solids) in solvent-based waxes
D1289-64	Nonvolatile matter (total solids) in water-emulsion waxes
D1342-57	Paraffin-type hydrocarbons in carnauba wax
D1792-66	Potential long-term removability properties of emulsion floor polishes
D2048-64T	Powdering of floor polishes
D1387-59	Saponification number (empirical) of natural waxes

TEXTILES

D2281-64T	Evaluation of wetting agents by skein test

PAINTS

D1955-61	Gel time of drying oils
D1849-63	Package stability of latex paint

BITUMEN

D1010-58	Asphalt emulsions for use as protective coatings for metal
D1167-65T	Asphalt-base emulsions for use as protective coatings for built-up roofs

GREASES

D217-65T	Cone penetration of lubricating grease
D1403-62	Cone penetration of lubricating grease; quarter-scale

D566-64	Dropping point of lubricating grease
D1742-64T	Oil separation from lubricating grease during storage
D1290-57	Sediment in water-emulsion waxes by centrifuge
D2047-64T	Static coefficient of friction of waxed floor surfaces
*D1796-62	Water and sediment in fuel oils by centrifuge
D1793-66	Water spotting of emulsion floor polishes

CLEANERS

D800-58	Metal cleaning compositions, industrial: chemical analysis
D1279-57	Metal cleaners: buffering action
D1281-57	Metal cleaners: rinsing properties
D1280-57	Metal cleaners, soak tank, total immersion corrosion test
D1374-57	Metal cleaners, aerated total immersion corrosion test
D930-49	Aluminum cleaners, water-soluble, total immersion corrosion test
D1567-66T	Detergent cleaners, evaluation of corrosive effects on certain porcelain enamels

References

1. Osipow, L. I., *Surface Chemistry: Theory and Industrial Applications*, pp. 18-21; Reinhold Publishing Corp., New York, 1962.
2. Orr, C., Jr., and Dallavalle, J. M., *Fine Particle Measurement*, p. 7; Macmillan Company, New York, 1959.
3. Montgomery, D. W., New Techniques of Particle-Size Analysis, *Rubber Age*, 759, February 1964.
4. Thacher, *J. Phys. Chem.*, 56:795 (1952).
5. Rosen, M. J., Recent Advantages in the Analysis of Surface-Active Agents by Non-Instrumental Methods, *J. Am. Oil Chemists Soc.*, 40:687, November, 1963.
6. Desmond, C. T. and Borden, W. T., Identification of Surface-Active Agents in Admixture by Thin-Layer Chromatography, *J. Am. Oil Chemists Soc.*, 41 (8):552 (1964).
7. Blank, E. W. and Kelly, W. E., *Soap Chem. Specialties*, 34 (1): 41 (1958).

chapter 5

TECHNIQUES OF EMULSIFICATION

Although new equipment and new surfactants are continually being developed, there are basically only two techniques of emulsification. In 1946, Cobb[1] succinctly defined the methods of emulsification as "brute force" and "persuasion".

Cobb may have been facetious, but nowhere in the field of emulsification is art more predominant than in the technique of producing the emulsion. An emulsion may be so formulated that a 10-year-old child with a wooden spoon can prepare it beautifully in the bathtub. More often than not an emulsion must be prepared "diplomatically" —gentle persuasion of the right combination of ingredients and "an artful kick" at just the right point.

Textbooks traditionally follow a staid approach and say that the two methods of emulsification are the *English* and the *Continental.* The use of these two methods in the preparation of an oil-in-water emulsion will be described. For a water-in-oil emulsion, just follow the same procedure using opposite phases (water for oil and vice-versa).

English Method

Mix small portions of the water and the surfactant(s) then add the remaining oil and water in successive, small portions.

Continental Method

Mix the oil and the surfactant(s) thoroughly and add the water to this paste.

Initial addition of water forms a water-in-oil emulsion which gradually thickens to a heavy paste (thick phase) with gradual addition of more water. Additional water or mechanical shear will invert the water-in-oil emulsion to the desired oil-in-water form.

The particular method for the best final product depends upon the the particular ingredients employed. The only general rule to guide the neophyte might be: Follow the Continental method unless you are in England.

"Brute force" (mechanical mixing and shear) can be added to either method at several points in the procedure. Some emulsions require massive doses of mechanical agitation of one sort or another once or several times during the blending of the emulsion. Generally, the emulsion-to-be may be subjected to mechanical agitation continuously during the blending, or at any of three stages.

Mechanical dispersion of the surfactant into one of the phases is desirable to obtain maximum effectiveness of the surfactant. If the surfactant is only slightly soluble in the phase chosen for dispersing it, mechanical agitation is usually necessary. Mechanical agitation will also aid the dispersion as the oil is added to the water (or vice-versa). The forming of the high-viscosity "thick phase" also provides an ideal time for mechanical agitation. The shear effects of mechanical agitation are multiplied by the effects of viscosity during the thick phase.

The time of mechanical agitation must be balanced by the type of agitation to be used. Four general types of mechanical force are usually applied to an emulsion.

1) *Beating* in the form of shaking or stirring is one of the most common forms of mechanical force. Dispersion is achieved through exposure to more or less direct contact with pulsating or rotating mechanical elements. The extent of agitation varies from a simple shaking or the use of a propeller stirrer in the laboratory, to the use of a colloid mill (characterized by a high-speed rotor moving inside a fixed stator with variable, but small, clearance).

2) *Pumping and injection* of one phase into another are the characteristics of the second type of mechanical force. The two phases are intermixed and dispersed by pressure-injection or -ejection, through the use of nozzles, pump casings, capillary tubes, or spring-loaded valves. The most common industrial application of this procedure is through simple pumping of the emulsion-to-be, or use of the homogenizer. Pressures in the order of 1500 to 8000 *psi* are achieved in the homogenizer as the fluid is pumped through a spring-loaded valve.

3) The internal frictional forces caused by greatly accelerated fluid motion, or *cavitation*, produce the third type of mechanical agitation. Particles in the emulsion are rearranged by the induction of periodic excitation. The ultrasonic emulsifying machines are an excellent example of this type of mechanical agitation.

4) *Mixing and interdiffusion* of particles in suspension resulting from the addition or withdrawal of heat can also be used; in fact, this type of mixing is often an unavoidable result of methods of agitation in which some of the mechanical energy is converted into heat. The mechanical shear resulting from the use of a colloid mill often adds heat to the mechanical shearing action.

The use of mechanical agitation often results in improved dispersion and smaller particle size. Mechanical dispersion is often necessary for maximum stability of the emulsion and stability in storage. Many skilled technologists, however, fail to realize that although a little agitation for a long time (or a lot of agitation for a short time) is beneficial, the over-use of agitation can reduce the quality of the emulsion. The rigid adherence to Aristotle's Golden Mean will result in the best products. Therein lies the secret of the *art* of emulsification.

Although mechanical agitation can improve an emulsion, it cannot make up for improperly chosen ingredients. Assuming that the ingredients have been chosen judiciously, the following practical hints on the technique of emulsification should be of help:

1. In most cases it is advantageous to add the internal phase slowly to all or part of the external phase. In special cases the preformed emulsion[2] may be used as a base for a new emulsion. (An emulsion can always be diluted with its continuous phase, but the stability may change.)
2. For O/W emulsions it is usual to put the emulsifying agent in the aqueous phase. When soap is used, it may be dissolved in the aqueous phase, or it may be formed *in situ* by dissolving the alkali in the water and the fatty acid in the oil.
   The latter method often gives better dispersions because the emulsion forms spontaneously as the two phases meet. Furthermore, less power is necessary for mixing in such a case. Certain emulsifying agents that are dispersible in both water and oils, e.g., diglycol laurate, may be dissolved in either the water, the oil, or both, depending on the type of emulsion desired.
3. In making O/W emulsions with soap, the oil and water are

alternately added into the concentrated aqueous soap solution. When a wax is being emulsified, it is usual to add the melted wax to a hot soap solution containing all of the water. When emulsifying a mineral oil with soaps that are soluble in it, a solution of the two is made and then stirred into the water.

4. Emulsions containing solids are best prepared by mixing the phases when both are at, or slightly above, the melting point of the solid. When an emulsion of a solid melting above 100°C is desired, the solid should first be melted with sufficient solvent or oil to bring the melting point below 100°C (e.g., rubber with naphtha or other hydrocarbons; synthetic resins with hydrocarbons or vegetable oils).
5. Incorporation of too much air causes extensive frothing, which is deterimental to the formation or stability of the emulsion.
6. Dilution of a concentrated emulsion should be accomplished slowly.[8]
7. Emulsions can be made with the help of an insoluble powder, such as clay, zinc dust, or yeast.[9] The quality of the emulsion depends on the fineness of the powder and its quantity, and on the presence of other constituents of the emulsion.
8. Mineral oil emulsions can be made without an emulsifying by agent shaking a small amount of oil with a large amount of water for 48 hours; or boiling a similar ratio under reflux for 30 hours; or by dissolving a similar ratio in alcohol and pouring the solution into water.[10]
9. For petroleum emulsions, the alkali is dissolved in water and the fatty acid in the petroleum. On mixing these solutions, soap is formed spontaneously and often gives a finer dispersion than other methods of mixing.
10. Bentonite, the most widely used insoluble powder, is dispersed properly by adding it to water, a little at a time, and allowing it to sink before adding the next portion. When all the bentonite has been added, allow it to soak for 1 or 2 hours, before stirring. Up to 7-8% of bentonite may be dispersed in this way. If a high-speed mixer is available, the bentonite can be added slowly while mixing, and dispersion can be effected in a few minutes. This method and longer mixing may yield dispersion up up to 17%.
11. Gum emulsifying agents should not be excessively diluted with water, because instability will result.[11]

References

1. Cobb, R. M. K., Section 2 in *Emulsion Technology*, Chemical Publishing Co., New York, p. 7 (1946).
2. British Patent 4505 (1912).
3. Lechler, British Patent 254701 (1925).
4. Billinghame, U.S.Patent 1700581 (1929).
5. Braun, British Patent 254012 (1925).
6. Robinson, U.S.Patent 1923888 (1933).
7. Neveu, U.S.Patent 1525409 (1925).
8. Brown, *Textile Colorist*, 57:516 (1935).
9. Bechhold, *Kolloid-Z.*, 28:6 (1921).
10. Lewis, *Kolloid-Z.*, 4:211 (1909).
11. Roon, *Ind. Eng. Chem.*, 9:156 (1917).

chapter 6

EMULSIFYING EQUIPMENT

The most important piece of equipment in the emulsifying plant is the device that effects the emulsification. The variety of available machinery can be arbitrarily divided into low-and high-shear devices. All of the devices mentioned are obtainable from a number of sources. Refinements and modifications have produced additional designs for agitators and mixer that are usually proprietary.

LOW-SHEAR EMULSIFYING EQUIPMENT

The low-shear devices include *propellers* and *ribbon mixers*. They are used either as the sole piece of equipment for emulsification or as blenders to disperse the ingredients of the emulsion before it is subjected to high-shear emulsification. These devices are primarily mixers. A satisfactory emulsion must be produced with little mechanical work for these low-shear devices to be effective. The basic, low-shear device is a power-driven agitator.

The operating conditions, that is, type of emulsion, concentration of the dispersed phase, additives, temperatures, and viscosity determine the type of agitator to be used.

Impellers

The results of processing are affected primarily by the type, size, speed, and performance of impellers, and their number and location. The geometry of the enclosure—baffles, dimensional ratios, and retention space—is of secondary importance. Another factor that plays a part in selecting an impeller is its reaction to changes in viscosity in terms of the power demanded. Finally, the available processing time may impose a limitation. Many emulsions, such as those containing colloidal matter, must be processed at moderate speed.

Many impellers are used singly or in series. The more common types are fabricated from a variety of materials and include the following:

Whipper:

An egg beater is a familiar example of the "whipper". A whipper, which is run at high speed, prepares the emulsion through the physical beating of the liquid mass. Two intermeshing grids that rotate in opposite directions are often used to supply the mechanical agitation.

Propeller:

The three-bladed marine-type propeller (Fig. 6:1) with a 45-degree

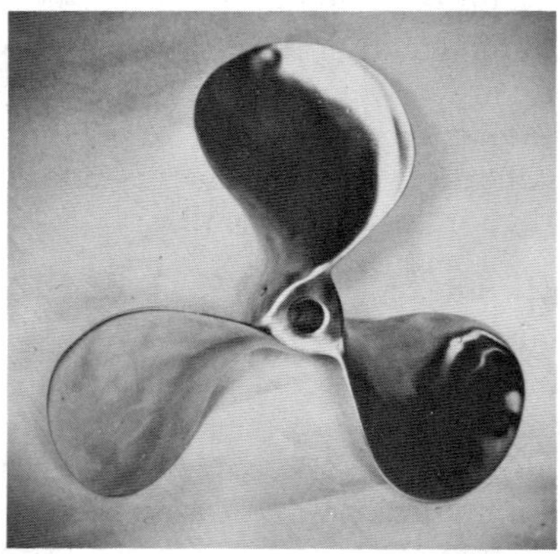

Fig. 6:1. THREE-BLADED MARINE-TYPE PROPELLER
(From *Mixing Equipment Company*, Rochester, New York)

pitch is the most versatile model for mixing materials with viscosities less than 3000 to 5000 centipoise. This propeller runs at speeds of 400 to 2000 *rpm*, and power requirements increase in proportion to the viscosity of the liquid. The rugged configuration of this propeller makes it especially suitable for use with abrasive or corrosive materials.

Because their mixing action is rapid, they may be used for both batch and continuous work. Their first cost is moderate. If they are mounted off-center and angularly they are excellent contactors of low viscosity immiscible liquids where not too fine a break-up is desired.[1]

The propeller type of agitator has undergone several significant modifications over the past few years. The range of viscosity that can be mixed with this type of agitator has been expanded through the use of impellers with a saw-toothed circumference. The *Cowles* design, using peripheral speeds to 6000 feet/minute, has been used for materials of 60,000 centipoise. The Cowles Hi-Shear Impeller is installed vertically, concentric with the tank (Fig. 6:2). The *Twin Daysolver* obtains

Fig. 6:2. SAW-TOOTH IMPELLER OF *Cowles* DISSOLVER
(From *Morehouse-Cowles*, Inc., Los Angeles Calif.)

increased shear by means of two counter-rotating spindles carrying perforated impellers with serrated circumferences (Fig. 6:3). The

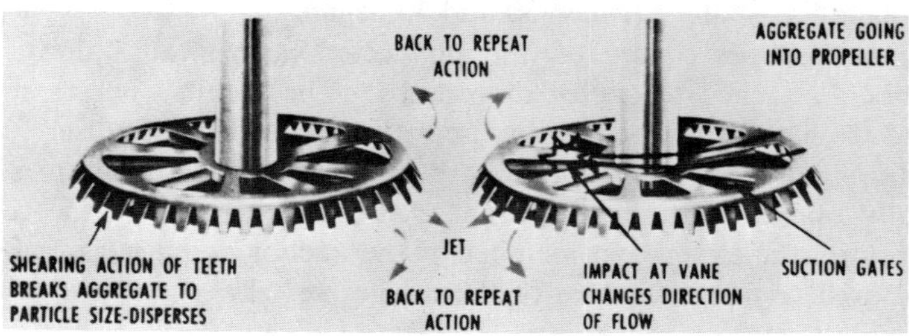

Fig. 6:3. COUNTER-ROTATING IMPELLERS OF DAYSOLVER
(From *J. H. Day Company*, Cincinnati, Ohio)

Barrington Disperser has a single shaft pointing down and an enclosed impeller. The impeller can be equipped with a short draft tube to avoid the formation of a vortex. The tube can be split and hinged, so that the halves can be swung out during operation.

Turbine impellers:

Turbines are classified as either *open* or *disk* type. The open type has flat or curved blades that are attached directly to the hub or shaft.

Turbine impellers are used for materials with viscosities of 50,000 to 200,000 centipoise. Their power requirements increase as the fifth power of the diameter.

In the *disk-type of turbine*, the blades are attached to a disk fitted to the hub. Four to six blades extend outward from the shaft of a radial, flat-blade turbine. The blades are placed vertically and at right angles to the horizontal. Thus, the radial turbine produces a flow that is essentially horizontal and radial. These turbines are primarily suitable for low tanks whose diameter exceeds their height.

Pitched-blade turbines usually have a 45-degree pitch, rotated to produce downward flow. They are used in tanks whose height exceeds the diameter. The range of viscosity that can be handled with a pitched-blade turbine is up to 20,000 centipoise, occasionally up to 100,000 centipoise.

In *curved-blade turbines*, the blades are usually arc-shaped segments of a circle of large diameter. The blades are usually in a vertical position, with the convex side in the direction of rotation; thus, the emulsion will roll off the blade during the mixing. Very viscous materials, including fibrous suspensions, can be handled with such impellers. The angle of back-slope (between a tangent to the blade at the periphery and a line connecting the center of the impeller and the tip of the blade) is usually between 30 and 60 degrees.

Gate agitators consist of interconnected horizontal and vertical members attached to one or more hubs. These units operate at low speeds to handle materials with viscosities to 10,000 centipoise.

Anchor or *horseshoe agitators* are similar to the gate agitators and usually operate with close tolerances at the inside wall of the vessel. Therefore, the lowermost member of the anchor agitator assembly is contoured to fit the bottom of the mixing vessel.

Mixers

Helical ribbon mixers have one or two helices on a vertical shaft. These helices move the emulsion near the of wall the vessel in a direction normal to the face of the blade. This mixer is used for pasty and sticky materials with viscosities of 10^4 to 10^8 centipoise. Operating speeds of the helical ribbon mixer are of the order of 10 rpm.

The *Cone-Vertical Mixer* (Atlantic Research Corp., Alexandria, Va. 22314) is a modification of the helical ribbon mixer. The container has the shape of twin intersecting cones in which tapering helices move close to the walls. The smooth walls facilitate downward motion caused by the downward push of the impellers. The product is extruded through a valve in the bottom of the mixer after the mixing operation is complete. Mixing is effected over a wide range of batch volumes because the geometric configuration of the batch and the flow pattern are always the same. The unit is jacketed for temperature control, and it can operate at positive or negative pressure. The Cone-Vertical Mixer is used for material from several hundred to several thousand poise.

Jets, orifices, turbulent-flow tubes, and pumps are excellent for the continuous blending of low-viscosity liquids. Where space is available for its installation, a turbulent-flow section of pipe 50 diameters in length is perhaps the simplest of all mixers. The designer should be assured, however, that the flow is turbulent. Jets and orifice columns will provide fairly coarse dispersions of immiscible liquids with economy, but fine dispersions may involve unattractively high pressure drop and power. These types operate best at one capacity, and the results obtained are variable when this is changed. The centrifugal pump is a good contactor, but it is difficult to control.[2]

Centrifugal pumps of standard design are used in combination with a recirculating tank to supply adequate intermixing of low-viscosity liquids. Some blending will take place within the pump in such a flow system.

A typical *continuous* blender assembly comprises premixing and mixing units, a main feed pump of the positive displacement type, and proportioners for additives, drives, and controls. Solids are conveniently added as premixed slurries.

The *Sigmund Mixer* consists of a mixing vessel and a recirculating pump. The vessel has an internal, agitating circuit and an external, recirculating circuit. The pump draws from a mixing vessel and discharges the fluid through a central vertical pipe and nozzle placed close to the bottom of the mixing vessel. This arrangement reverses the flow in the mixing tank and causes intense turbulence. The various components are introduced to the outer circuit through a common header. The operation is continuous through the continuous metered addition of ingredients at the header and the removal of a like amount of finished product.

A smaller volume and a lower time of retention in a continuous-operation system is characteristic of the *Ashbrook Vortex Mixer*. The main stream of fluid moves down a central pipe and entrains admixtures by an injecting action. The configuration and dimensions of the unit must be carefully designed to accomodate the properties and quantities of the materials to be processed.

The *Premier Dispersator* is a modification of a centrifugal pump with external recirculation. It consists of the impeller of a pump without its stator. The impeller is open at the bottom and rotates at high speeds. The volume of the container is relatively small compared to the throughput of the impeller, and the rate of recirculation is accordingly high.

HIGH-SHEAR EMULSIFYING EQUIPMENT

High-shear devices have been developed specifically for the operations of emulsification and dispersion. Common high-shear devices effect dispersion in one of three basic ways: with rotor-stator devices, pressurized fluid devices, vibrational devices.

Rotor-stator Devices

In rotor-stator devices, the emulsion passes through the very narrow annular space between a rotor and a stator (e.g., colloid mills).

Turbines, turbodispersers, and other emulsifying devices characterized by a high-speed rotor and matching stationary stator are included in the classification of colloid mills.

Colloid mills are common, high-shear emulsifying machines. They are characterized by rotors with peripheral speeds of about 10,000 feet / minute. Centrifugal action moves the feed to the periphery of the rotor and back through an adjustable gap between the rotor and the the stator. The gap may be set at 0.005 to 0.05 inches. Colloid mills are used with fluid viscosity up to 5000 centipoise, and their rate of output is inversely proportional to the viscosity of the emulsion. A broad line of colloid mills for both laboratory and production plant is made by the Manton-Gaulin Company of Everett, Mass. 02149. These units illustrate the operation of a colloid mill (Fig. 6:4).

Premier Mill Corporation produces a line of colloid mills with average rate of flow of 30 to 1500 gallons an hour. The floor space for such units varies from 12×25 feet to 32×32 feet for the larger models. The units normally are fed vertically by gravity flow. *Charlotte Colloid Mills* are available in capacities from small, pilot-plant models to production units capable of emulsifying 7000 gallons an hour. The material being processed is either gravity-or force-fed to the space between a stator and a 3600-*rpm* rotor. The grooved rotor and stator emulsify the material by means of hydraulic shear, violent turbulence, and the impact of the eddying currents caused by the grooved rotor and stator.

The *Tri-Homo Homogenizer* is characterized by a three-faced rotor.

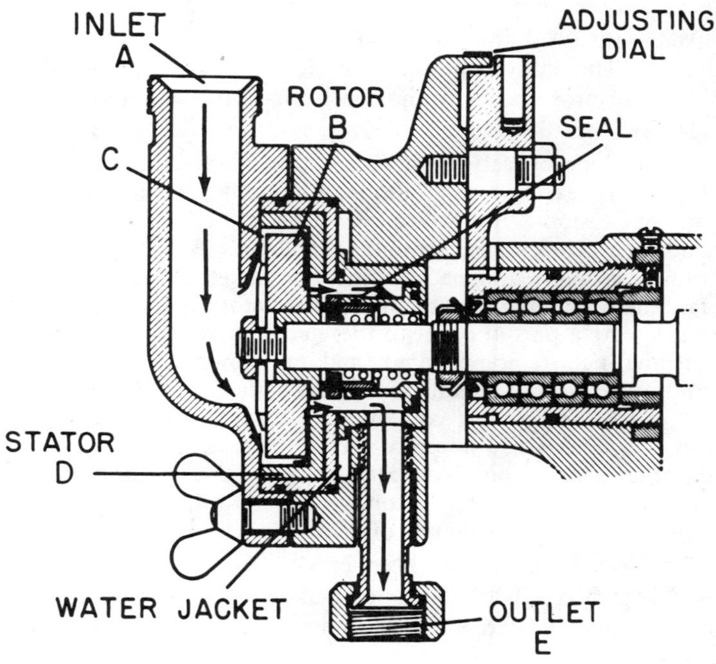

Fig. 6:4. CROSS SECTION OF COLLOID MILL
(From *Manton-Gaulin Manufacturing* Co., Inc., Everett, Massachusetts)

This colloid mill is also available with a variety of rotors designed for the specific emulsifying operation and degree of shear desired (Tri-Homo Division, Patterson Industries, East Liverpool, Ohio 43920).

The Gifford-Wood Company(Hudson,N.Y.12534). produces a series of *Eppenbach colloid mills* that are suitable for either batch or continuous operation. In many models the entire rotor and stator assembly is submerged in the emulsion. The Eppenbach design embodies a substantial stator with several cylindrical passages. The four-bladed rotor operates at high speeds and is mounted to maintain a close and controlled clearance from the stator. Eppenbach—design mixers are also available in modifications for both laboratory and production high-shear mixing.

The *Barrington Convertible Jet Mixer* is characterized by a convertible feature that allows use as either an open impeller mixer or a closed impeller, colloid-mill type of mixer. The tolerances between the star-shaped stator on the central shaft and the pitched-vane rotor can be adjusted. The unit can be converted to a closed design by attaching a cylindrical stator sleeve around the rotor-stator assembly (Barrington Industries, Inc., 185 Union Ave., Providence, R.I. 02909).

> A turbine operating within a stator will excel all other types (of mixers) except a colloid mill and a homogenizer in producing fine dispersions. Furthermore, they are the fastest of all types, and their power consumption is moderate. Because of the ability of a properly designed turbine-type mixer to approximate instantaneous mixing, it is the most satisfactory type for continuous work. As a partial offset to these advantages the structure of a turbine is more complicated (making it more difficult to clean) and its first cost is usually somewhat higher.[3]
>
> The turbodisperser consists of a centrifugal turbine impeller which rotates with a screen or perforated plate interposed between impeller and stationary deflecting blades. ...... The high degree of shear, the extruding action, and the high flow, all contribute toward the accomplishment of dispersions.[4]

The *Premier Dispersator* (Premier Mill Corporation, 224 Fifth Ave., New York, N.Y. 10001) is a modified semiclosed impeller-mixer. A slotted cylindrical head with internal baffles rotates at high speed. The emulsion is drawn in through the bottom of the unit and passes through the cylindrical head before being thrown out of the head by centrifugal action through small slots. The product is sheared hydraulically as it passes through the slots and is sheared by the "blades" of material emerging from the cylinder and knifing into the slower-moving liquid mass outside the head. Both batch and continuous designs are available.

The *Waukesha shear pump* uses a series of interchangeable rotors and stators equipped with slots and drillings in different arrangements to control time of retention and particle size. This shear pump is often used in dispersng operations.

The *Versator disperser* (Fig. 6:5) employs a bowl-shaped disc that rotates at speeds up to 3400 *rpm* in a chamber. The emulsion is introduced to the center of the discs and centrifugal force drives the material away from the center, with decreasing thickness of the film and increasing turbulence. The material is subjected to high-shear forces as it moves across the disc, causing effective homogenizing and dispersing action (The Cornell Machine Company, 45 Brown Ave., Springfield, N.J. 97081).

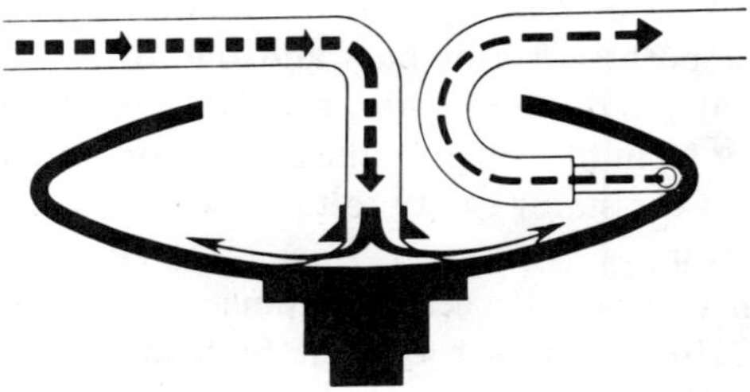

Fig. 6:5. FLOW OF FLUID IN VERSATOR EMULSIFYING MACHINE
(From *Cornell Machine Company*, Springfield, New Jersey)

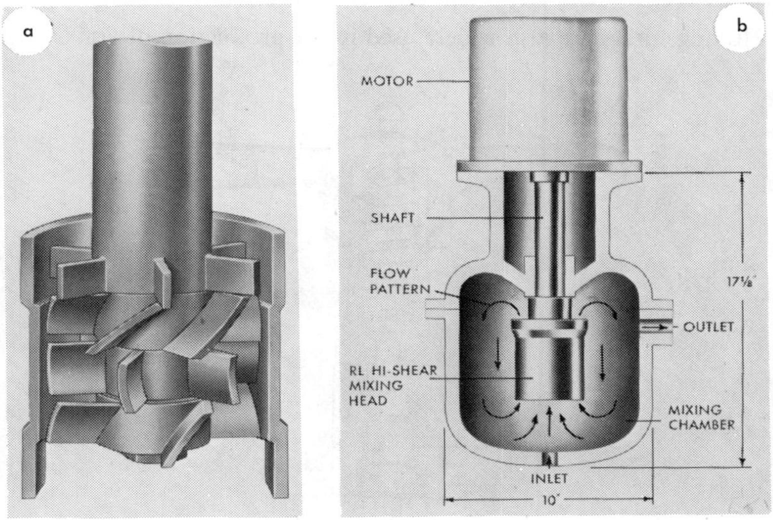

Fig. 6:6. DETALS OF EMULSIFYING MACHINES
a) Shear head, showing twin rotors and dual stators; *b*) Recirculating In-Line Mixer. (From *Gabb Special Products, Inc.*, Windsor Locks, Connecticut)

The *Gabb shear flow mixer* uses two impellers and dual stators. It has been used for materials of viscosity between 10 and 100,000 centipoise (Figure 6:6a). Another type of closed impeller-mixer is derived from the squirrel cage disintegrator or shredder. It has stationary systems with concentric stator bars and one rotor in between. The rotor operates at several thousand *rpm*. The ingredients are fed to the mixing chamber from the center of the rear stator, and pass to the outer circumference of the stator cavity to the discharge nozzle at the center of the front stator.

The *Turbon* (Turbon Process Equipment, Inc., 44 East 29th St., New York, 10016) is another type of high-speed centrifugal mixer.

The *In-Line Mixer* designs are attractive where large volumes require only brief mixing times.

Lightning and *Gabb In-Line Mixers* (Fig. 6:6b) have less shear than the Eppenbach design.

Pressurized Fluid Devices

In pressurized fluid devices, the emulsion is expelled at high pressure through the annular space between a valve and its seat (e.g., homogenizers).

Homogenizers are also widely used in the production of emulsions.

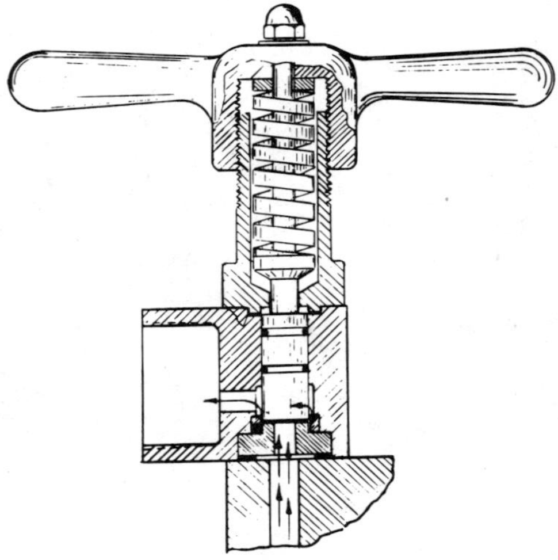

Fig. 6:7. SINGLE-STAGE PLUG-LINE HOMOGENIZER
(From *Manton-Gaulin Manufacturing* Co., Inc., Everett, Massachusetts)

A homogenizer contains a positive displacement pump that forces the emulsion through an annular slot between a valve and its seat. The valve is spring-loaded, and the load can be adjusted. (Fig. 6:7). The acceleration and turbulence that result from the shear of passing through a narrow passage and sudden reduction in pressure produce particle sizes of less than one micron. Usually the emulsion should be less than 400 centipoise for maximum effectiveness. Less heat is produced in the homogenizer than in the colloid mill for emulsions of the same resultant particle size.

The following remarks help to differentiate between the use of a homogenizer and a colloid mill in the manufacture of emulsions of fine particle size:

> The first consideration in determining whether to use a colloid mill or a homogenizer would be quality. If the emulsion or dispersion requires the finest possible particle size, then the homogenizer should be used. If something less than the ultimate is satisfactory, then the mill may be considered. Like all general statements there are exceptions. Emulsions in which the dispersed phase is at the highest possible concentration, complicated by a low percentage of emulsifier as in mayonnaise, must be processed in the mill. If the homogenizer were used, the pressure acting on the dispersed phase and surfactant would cause more emulsifier to be combined with the dispersed phase than will occur in the colloid mill with its lower pressure. Fortunately, most emulsions do not have a formulation limitation and the ability of the homogenizer to combine more stabilizer to the dispersed phase becomes an asset. In the case of pigments or other solids in liquid dispersions, the homogenizer will show a marked superiority over any other type of equipment when the solid particle size is one micron or less. For solids having an ultimate particle size in excess of one micron, the mill should be considered. The mill may also be preferred if the batches are small and of varying colors.
>
> Resin emulsions of high viscosity such as the *hot melt type* would require the colloid mill. The high internal viscosity of this type of system prevents the efficient use of the high velocity shear generated in the homogenizing valve. In many instances the product will stream across the valve surface in the form of a thin ribbon without breaking and immediately reform into large globules on emerging. In the colloid mill the same general action takes place, except the rotation of the rotor breaks up the ribbon and reduction in particle size is accomplished.[15]

In addition, the homogenizer and colloid mill were characterized as[6]:

Homogenizer	*Colloid Mill*
Low mechanical speed	High mechanical speed
High material speed	Low material speed
Constant capacity regardless of pressure or viscosity	Varying capacity depending on gap-setting and viscosity
Low rise in temperature of product	High rise in temperature of product
Unlimited feed and back pressure	Limited feed and back pressure

Several other homogenizers are available besides the Manton-Gaulin models. *Hydropulse* homogenizer is designed specifically for the sanitary processing emulsions, or handling abrasive and corrosive liquids. The machine is designed to facilitate ease of cleaning and sterilization. Hydropulse homogenizers are also available for a wide variety of applications (Scott and Williams, Inc., Laconia, N.H., 03246).

The *Logeman Laboratory Homogenizer* is a bench-top unit capable of processing samples from 30 cc to 8 gallons an hour. Production models with capacities of 25 to 2500 gallons an hour are also available (C. W. Logeman Company, 633 Bergen Street, Brooklyn, N.Y. 11238).

Vibrational Devices

In vibrational devices, a vibrating element causes cavitation and results in particulate dispersion (e.g., ultrasonic emulsifiers).

When a vibrating element submerged in a liquid moves very rapidly, the liquid around it cannot follow the movement of the element because of inertia in the liquid phase. The discontinuity of motion between liquid and vibrator causes local cavitation leading to emulsification of many systems. This principle of electrical emulsification is the basis for the following methods of emulsification:

Piezoelectric:

Certain crystals contract when placed in an electric field. If an alternating field of the same frequency as the natural mode of vibration of the crystal is applied across the face of such a crystal, high-amplitude vibrations result in emulsification energy. The *Branson Sonifier* is a piezoelectric sonic emulsifier designed for research and production. The smallest unit is capable of emulsifying up to 20 gallons an hour with the addition of a attachment for continuous flow.

Electromagnetic:

High-intensity sound waves can be generated by exciting alumina rods at their resonant frequency by using the induction principle employed in moving-coil speakers.

Magnetostriction:

Certain ferromagnetic materials change length when placed in a magnetic field. The small change in length is independent of the direction of the field but varies with the strength of the field. An alternating magnetic field with the same frequency as the natural mode of vibration of a nickel rod produces resonance that results in the generation of powerful waves. As a magnetostriction vibrator carries no electrical current, it can be operated directly in a conducting fluid.

Ultrasonic:

The liquid being processed is delivered at high velocity through a jet which directs it, in a thin flat stream, across a steel blade which is thus caused to vibrate. (Fig. 6:8). Sound energy is generated by the flow of the product itself and not by any external source of energy. A mixture of frequencies is generated inducing an extremely high level of cavitation within the stream of liquid.

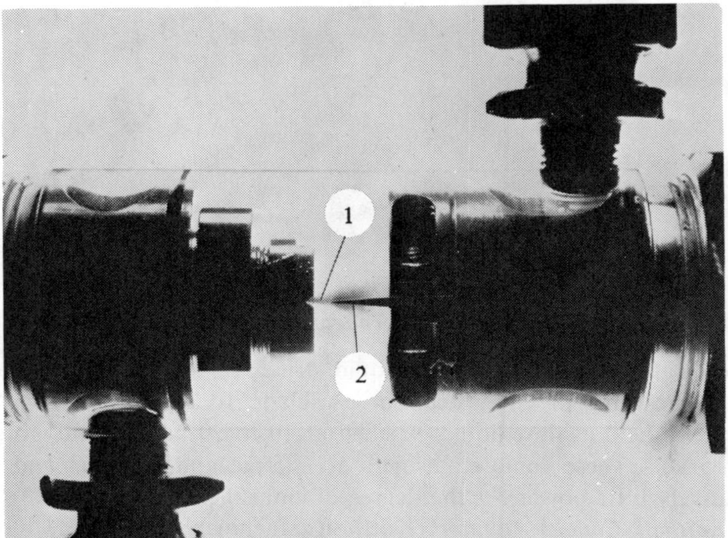

Fig. 6:8. Detail of Ultrasonic Emulsifying Machine
1) Jet of emulsion leaving orifice; 2) Vibrating blade (From *Sonic Engineering Corporation*, Norwalk, Connecticut)

Of these methods, only the last has achieved commercial significance at the present time. In the production of emulsions, Pohlman adapted and used the principle of acoustic waves generated by a whistle in an organ pipe.[9] A review of this method of emulsification was presented by McCarthy.[8]

The *Rapisonic Acoustical Emulsifier* expells liquid through an adjustable slit in a flat orifice against a stationary blade (Fig. 6:9).

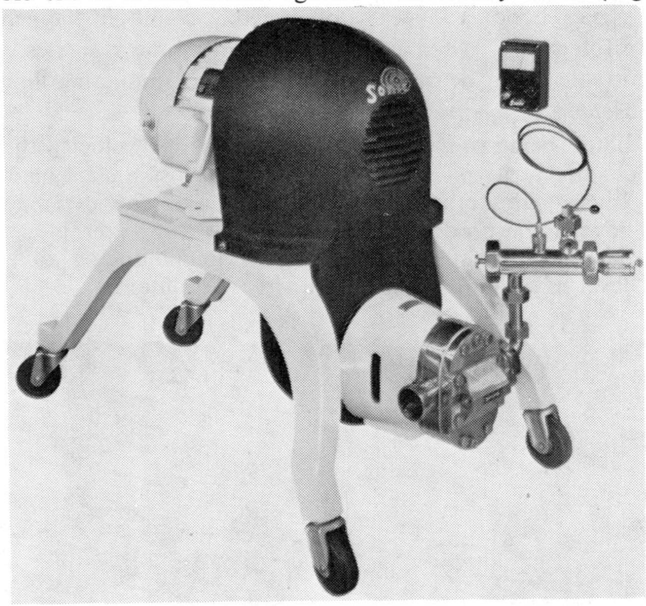

Fig. 6:9. SONIC MODEL OF EMULSIFYING MACHINE
(From *Sonic Engineering Corporation*, Norwalk, Connecticut)

The resulting cavitation as the liquid strikes the vibrating blade provides the force of emulsification. A blade 10 mm away from the orifice is held against liquid expelled at 10 to 20 kg/cm^2 and vibrated at 15 kc. These sonic emulsifiers are simple and compact and use relatively little power for the degree of emulsification obtained.

Kotthoff Colloid Agitator (Kotthoff Machinenbau GmbH, Kilchbery—Zurich, Switzerland) uses acoustical energy to achieve dispersion in particle size. The sound waves are reflected from the stator space to produce maximum cavitation and emulsification. Emulsions up to 8000 centipoise can be produced on either the laboratory or the production scale with the aid of this machine.

LABORATORY EMULSIFYING EQUIPMENT

As most emulsions are prepared in batches, the techniques followed in the laboratory and in the plant are similar. Laboratory equipment, although sometimes simulating production equipment on a small scale, is often unique. The simplest laboratory apparatus consists of a small closed container and a moderately strong arm. Emulsions are often made by shaking a small jar (usually 4- to 8-oz., two-thirds full) by hand. Manual shaking is not effective if the emulsion is very viscous or if a great deal of shear is necessary to provide a stable product.

Intermittent shaking may be more effective than equally violent uninterrupted agitation. For instance, the following:

Benzene	60% (Vol.)
Sodium Oleate	1%
Water	39%

requires 750 shakes and takes taking 4.2 minutes to emulsify the mixture. If the same mixture is shaken twice, allowed to rest for 30 seconds, and shaken three times, a comparable emulsion is formed in less than one-quarter of the time.

The mortar and pestle are often used to prepare crude emulsions, especially in a pharmacy. This equipment will provide emulsions of quality similar to those shaken by hand. This method is generally more effective for forming thick emulsions for reducing the particle size of solid materials that are to be emulsified.

The *Mulsichurn* is used in the laboratory to prepare test emulsions in small, contaminant-free quantities. A perforated dasher-piston is moved by hand in a solid sleeve. The liquid is forced through the orifices in the dasher at velocities of 300 to 800 inches a second, and emulsification is completed after 30 to 50 strokes.

Emulsions can also be prepared on a three-roll mill. The three-roll mill is often used to create pigment dispersions of fine particle size, and may be used for emulsification also. Each of the three rolls rotates faster than the preceding one. The emulsion is placed on the slowest roll and it transfers successively to the faster rolls in turn. The final product is removed from the last (fastest rotating) roll by a doctor blade. Thus the emulsion is sheared and its particle size reduced as it passes between two rolls of different speeds.

Several variations of propeller-type mixers are available for the laboratory preparation of emulsions. The propeller mixer is probably the most common piece of equipment used for emulsification. The size, shape, and relative position of the agitator and container, as well as the speed of the agitator, determine the quality of the resultant emulsion. The greatest degree of agitation without causing excessive entrainment of air will generally produce the best emulsion. The laboratory propeller agitator is usually powered by an electric or air motor of variable speed. Rotational speeds with these units can be varied from 50 to 2000 *rpm*. Explosion-proof electrical or air-powered units are recommended for use in potentially flammable atmospheres. A 1/10 hp counter-rotating mixer has been developed by Brookfield to mix materials and to avoid formation of a vortex. The counter-rotating propellers on two concentric shafts reduce entrainment of air and the risk of "slopover".

Homogenizers are frequently used to prepare emulsions of ingredients that are difficult to emulsify, or to obtain dispersions of small particle-size. A manual homogenizer is available for preparation of one- or two-ounce laboratory samples (Figure 6:10). Electric-

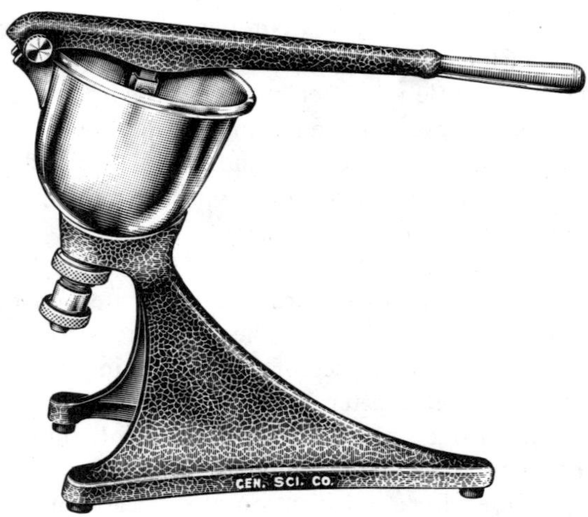

Fig. 6:10. Manually Operated Homogenizer for Laboratory (From *Central Scientific Company*)

ally powered homogenizers are also available for the preparation of larger samples.

Small production-type colloid mills are often used in laboratories for preparing emulsions of fine particle size. The clearance between the rotor and stator can be adjusted to vary the shearing stress on the emulsion.

Ultrasonic emulsifiers are available in laboratory sizes for the production emulsions of fine particle size.

References

1. PERRY, J. H., editor, *Chemical Engineers Handbook*, (3rd Ed); McGraw Hill Book Co., Inc., New York, 1950.
2. *Ibid.*, p. 1216.
3. *Ibid.*, p. 1215.
4. *Ibid.*, p. 1211.
5. *What Is The Difference between a Homogenizer and a Colloid Mill?;* A company publication, Manton-Gaulin Manufacturing Co., Inc., Everett, Massachusetts.
6. *Ibid.*
7. Ultrasonic Homogenizers, *Chemical Products and Aerosol News*, p. 36, May 1963.
8. POHLMAN, R., *Z. Angew. Phys.*, 4:181 (September 1948); *Phys. Abstr.*, (A) 53 (1950).
   — *Technik*, 3:465 (November 1948); *Phys. Abstr.* 52 (1949).
9. McCARTHY, W. W., Ultrasonic Emulsification, *Drug and Cosmetic Industry*, 94:821-4, June 1964.

chapter 7

EMULSION PLANTS AND PRODUCTION MACHINERY

This section is designed both to give the laboratory neophyte an introduction to the production aspects of emulsification and to present some of the special considerations that must be remembered by the design engineer in building a unit specifically for the production of emulsions. It is not the purpose of this chapter to teach plant design, heat transfer, mass transfer, stoichiometry, or chemical engineering cost estimation. The references give the detailed information needed on these subjects in order to design and build an emulsion plant.

The equipment. time, and energy required for the process of emulsification are dependent on:

1) The characteristics of the materials to be used. (As emulsification is primarily a flow process, the rheological characteristics of the materials are of prime importance).
2) The mode of operation (batch, semi-continuous, or continuous operation; temperature, pressure, agitation, addition procedures, etc.).
3) The characteristics of the equipment (area of phase-interchange, arrangement and distribution of fluid paths, amount of flow, shear, and heat transfer, co-ordination of components and auxiliary equipment).

The steps in the development of a plant and the over-all planning schedule will be established before turning to the details of the emuls--fication plant itself.

DEVELOPMENT OF PLANT

In order to produce emulsions economically on a large scale, several transitions from the laboratory to full production are necessary. In an effort to speed a product to the market, several steps of this scale-up procedure may be omitted. The risk involved in eliminating any step should be balanced against the cost (in terms of both money and time) of carrying out that step.

Preliminary Laboratory Work

The development of a new emulsion begins in the laboratory. Once the desired product is created in a small laboratory batch, a repeat of the identical formulation should be repeated several times. These duplicate samples are then tested to verify that the emulsion and its results are reproducible. From these and subesequent laboratory batches the chemical nature and basic physical properties of the emulsion can be characterized. The composition and a method of production are specified. Some of the critical variables in formulation and operation are identified, and the allowable limits are tentatively established. The final emulsion is tested to verify content and efficacy. Glass laboratory ware and small laboratory devices for shearing and mixing are used. Particle-size distribution, relative densitives of the phases, abrasive or corrosive properties, and flow characteristics are roughly established.

After the initial laboratory work has been completed and the formulation and allowable variations in it have been established, the scale of production of the emulsion is increased. Larger batches are made, using scaled-down models of anticipated production equipment. The rates of heating and cooling and their effect on the final emulsion are evaluated. Other variations in the process are evaluated, such as the order of adding components and the use of alternate suppliers of components. The work is still probably a small-scale operation in the laboratory. The tests that are necessary and sufficient to indicate an acceptable product should be in an advanced state of development by this time. These tests will be used for quality control in the eventual production of the emulsion.

Transfer to Pilot Plant

Pilot plant equipment is then used to supply data on power-consumption. Small sizes of standard equipment are used to produce the emulsion as it would be produced eventually in the production plant. The best process for manufacturing the emulsion is determined, and the control limits on all variables in formulation and in process and are established. Often the product that is made in this pilot plant will be used for the initial testing of the product in a limited market. Although this stage is not the last for a large producer of the emulsion, the supplier of an emulsion to a limited market may find that his development and production need go no further.

Semi-works Stage

Semi-works production, intermediate between pilot plant and full scale production, is important whenever the project requires considerable investment and the product or process is novel. The semi-works plant is, as nearly as possible, a complete prototype of the final production plant at reduced-scale. The product of this plant may supply the needs for regional marketing, or may be used to introduce the emulsion to the market on a national scale.

Production equipment finally evolves from this long and arduous procedure. The equipment, the process, and frequently even the emulsion itself, may bear little resemblance to the original laboratory sample. The entire system may have gone through several "back to the drawing board" modifications. Before the toast of victory is raised, however, the production plant must be running smoothly, and the tests for acceptance on all the equipment must have been satisfactory.

The typical layout of a plant for production of asphalt emulsions is shown in Figure 7:1.

Overall Planning Schedule

1. Select mode of operation based upon the rate of production and the expected market demand.

   The mode of operation may be *batch*, *semi-continuous*, or *continuous; single-* or *multi-stage; hand-controlled* or *automatic*.

   Batch operation is indicated where the volume to be processed is small, production is intermittent, or where different types and grades of products are in demand. Constant supervision is necessary to maintain quality of product and to avoid the hazards of contamination due to change-overs and shut-downs. As batch equipment is usually highly versatile, it is used in most manufacturing of emulsions.

   A *semi-continuous* operation may form an intermediate step in the expansion of a batch plant.

   Continuous operation often permits a savings in labor, heat, power, and raw materials. It is applicable even to complex production methods if the equipment is properly designed and operated. However, the production rate must be large enough

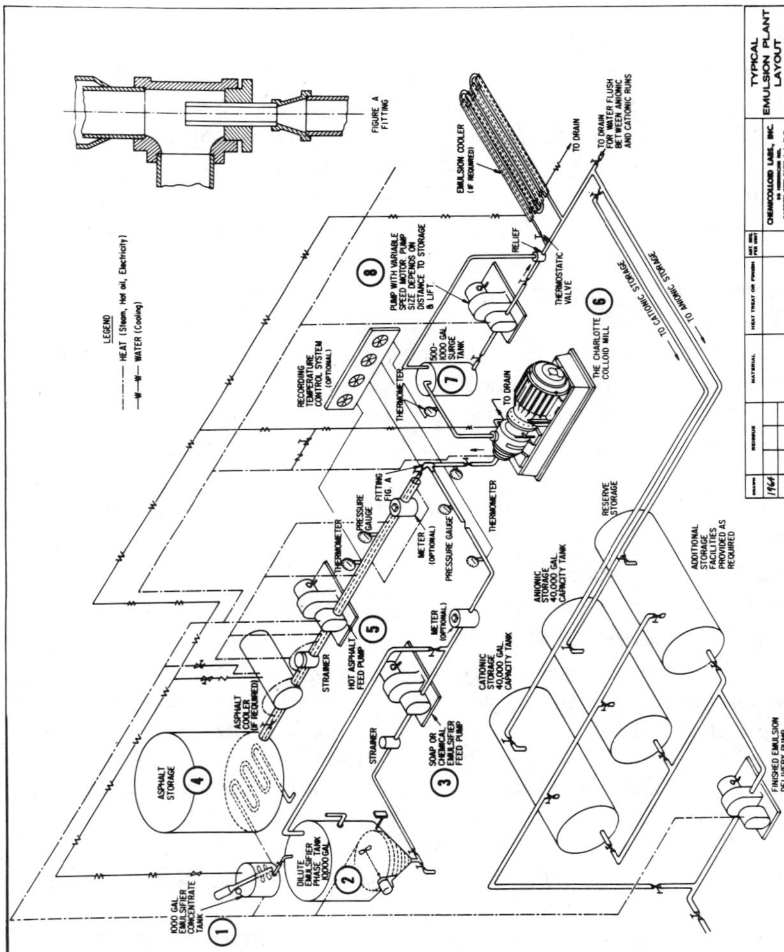

Fig. 7:1. LAYOUT OF PLANT FOR ASPHALT EMULSIONS (From *Chemicolloid Labs, Inc.*, Garden City, New York)

to compensate for the high initial cost of instrumentation and controls.

Multi-stage operation generally improves process output and may improve the quality of product if volumes are sufficiently large.

2. Draw a tentative flowsheet.

The relative complexity of the operation will be reflected in the layout of the plant.

3. Determine operational pressure and temperature levels and stream mass ratios.

Data can often be obtained from laboratory or plant records. Most emulsions are produced at atmospheric pressure. Operations under pressure, however, can speed reactions when necessary, but this requires closed containers. Applications of reduced pressure lowers heat requirements, facilitates deaeration, and prevents scorching. Relevant data can often be obtained from plant records to aid in design of plant. Chemical effects (reactions, side reactions, by-products) and physicochemical phenomena (heats of reaction, equilibria, sorption effects, corrosive action) must be considered. High temperatures resulting from these effects may cause decomposition or separation of the product.

4. Determine methods of heating and cooling.

Direct heating by oil or gas-firing is simplest, but hazardous and hard to control. Indirect heating and cooling requires jacketed equipment; the carrier (water, brine, steam, oil, molten salts) is selected to suit the temperatures involved. Most emulsion plants use steam, hot water, or a heat-transfer fluid (e.g. Dowtherm).

5. Size piping for product and auxiliary fluids.

Label individual streams with data on flow, temperatures, pressures, compositions. Minimum size of pipe for latex and similar materials is 2 inches. For multi-stage equipment establish levels of material in piece of equipment.

6. Establish over-all dimensions for processing units.

Standard equipment will suffice where viscosity is low and no accurate proportioning need be maintained. Space must be be allowed for drives, controls, replacement of tubes, and inspection. Establish the optimum combination of size and shape of vessels and capacities of impellers in agitators. If homogeneity of the product depends on time, size becomes critical. Observe requirements for retention-time, especially in multi-stage operation.

In the manufacture of lubricants and greases, kettles of very large capacities are often employed, but these require a carefully designed agitator. Packaging machinery requires considerable floor space.

7. Handling of materials.

If sufficient head is available, gravity-feed is sometimes sufficient. It may be satisfactory for moderate rates of processing and low pressures, especially if fluctuations in head are not objectionable. Solids may require gravity or other conveyors.

Where an adequate supply of compressed air is available, blowcases or *montejus* may be practical. These are closed vessels containing the batch to be processed. Air-pressure applied at the top forces the material to the emulsifier. This method is especially useful when the nature of the feed makes pumping difficult or expensive. The blowcase must be located fairly close to the source of the material to be handled, and so placed that it can be filled by gravity. The safety factors of possible corrosive or abrasive action must be considered.

8. List pumps, compressors, motors.

In vacuum operation, air pumps or steam ejectors will be required.

Electric drives predominate; *NEMA C flange motors* are typical. Enclosures for motors depend on local conditions—splash-protected, totally enclosed, explosion-proof, or equipped with special insulation. Explosion-proof designs should be specified if flammable solvents are used in the area. In hazardous locations, air motors or steam turbine drives may be preferable. The speeds of motors follow AGMA standards; speed is varied by interchanging pulleys or stepless transmissions.

9. Provide required accessories.

Various accessories are required to control flow, such as gates, valves and their actuators, metal hose and connectors, piping, spouts, line-diverters, air locks, collectors, separators, and check valves. Other accessories facilitate the supervision of the process, such as access ports, view ports, sight tubes, and liquid-level gauges.

10. Instrumentation and controls.

Especially in fully automated installations, care must be exercised to establish control points properly—for degrees of agitation, rates of heating and cooling, metering of ingredients, etc. Type and locations have to be determined for all instruments and their actuators and safety devices.

For complicated installations, models help in the arrangement of piping and control points. When all controls are properly installed and adjusted, constant quality of product must be assured without further attention from the operator. Most emulsions are produced with the minimum of instrumentation, but adequate control must be specified to ensure a consistent product.

11. Housing and shelters.

Large and rugged equipment may permit outdoor operation in most climates. Plot plans and elevations are used to determine supply points and arrange for loading and weighing equipment. Attention must be given to slopes, drainage, and hazards from fire and pollution.

12. Utilities.

As a rule, electricity, water, and steam will probably be needed. In large installations, recovery of heat (heat pump system) may be worth considering. Especially for high rates of production, balances in material and in heat should be verified. Needs for stand-by equipment and demands for overload must be taken into account. Utilities may be generated or purchased.

13. Safety.

In automatic installations, a fail-safe condition must be established when necessary to comply with local codes, insurance regulations, and general safety rules. All equipment must be explosion-proof. Main cutoff switches should be located in a separate area. Equipment likely to develop static electricity must be grounded. Tank cars, liquid cascades, dust bins and containers of solvents should be grounded.

14. Final phases.

Draw up foundations and steel-work in accordance with plot plan, soil conditions, etc. Write specifications and correlate contracts.

As a rule, the equipment in a process plant will have to be purchased from several manufacturers. Proper synchronization and careful planning of the components to assure the proper flow of raw materials and product are thus required.

Modification of "standard" equipment may become necessary, and a built-in degree of flexibility is desirable throughout the entire production line, without incurring complication in design or in manipulation.

Flexibility, however, should not be provided to the detriment of large design or layout considerations.

MODES OF OPERATION

Batch Process

In a *batch operation*, measured quantities of the ingredients are processed until the desired results have been achieved. After the unit operation is completed, the batch is discharged to packaging or further processing. The rate of production can be adjusted at will by the size of the batch or number of batches handled in a unit of time. In blending operations, a single vessel and agitator are often sufficient. Equipment costs can be held at a minimum. Operating costs are relatively high, but they can be reduce by automatic cycling.

In designing equipment for a controlled batch process, required data comprise:

1) Volume of batch
2) Time allowed for mixing
3) Sequence of adding ingredients; if the corresponding viscosities vary appreciably, the lighter components are charged first. Agitation is then started, and the heavier components are added gradually.
4) Expected variations in product characteristics, especially for non-Newtonian materials.

In Figure 7:2, dry raw materials and a premixed slurry are supplied to a continuous horizontal ribbon blender (4) by two proportioning scales (1, 12) and a proportioning pump (2). The relatively dry granulate still carries oversize particles which are reduced in the fluid energy mill and interdispersed. The horizontal kneader (14) to which pump (15) supplies additional semi-fluid ingredients exerts a plasticizing action The globules of the product, still irregular in size, are forced through a homogenizer (17). To further improve

---

1, 12	Dry solids feeder scales	10	Cyclone separator
2, 15, 19	Proportioning pumps	11	Collector
3, 16	Premix tanks	13	Kneader drive
4	Continuous mixer	14	Continuous kneader
5	Feeder	17	Homogenizer
6	Air lock	18	Control board
7	Impact mill	20	High-frequency emulsifier
8	Oil-free compressor	21	To packaging
9	Inert gas storage		

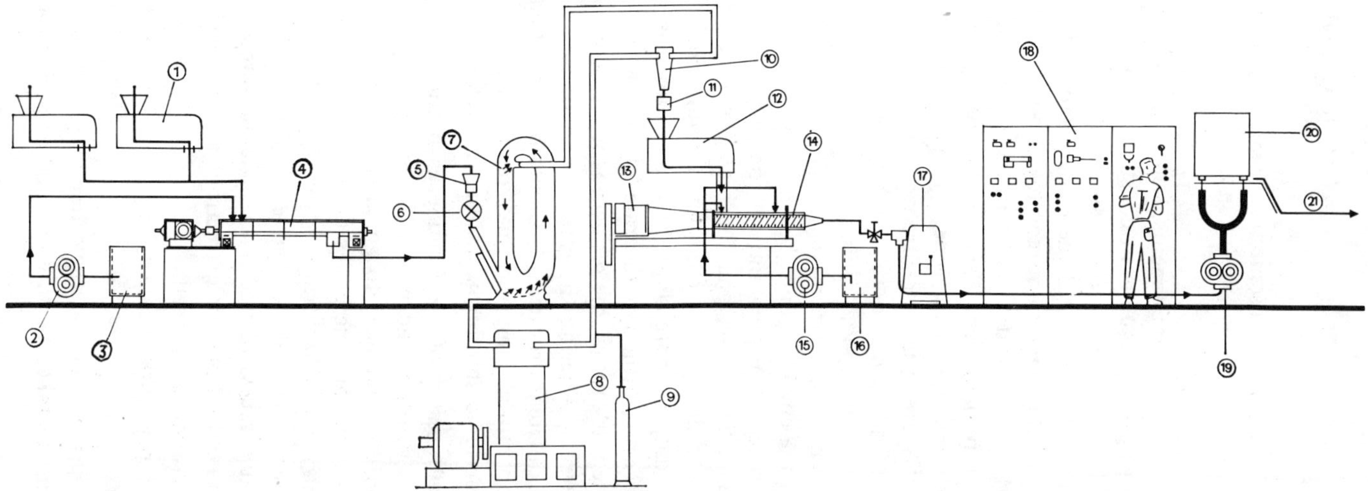

Fig. 7:2. Plant for Continuous Blending and Homegenizing. *From G. Mosimann & Co.* (see p. 106 for key)

the fineness of the dispersion, ultrasonic treatment is applied in unit 20. The installation requires only one attendant.

Continuous Process

In a *continuous process*, the components, which may be premixed, are fed to the dispersing unit or units in constant proportions (Fig. 7:2). The retention time can be controlled by the flow from the discharge nozzle and/or by the settings of the feeders and proportioners of the ingredients. If the operation requires heating or cooling, provisions for the transfer of heat to or from the process, and sufficient insulation, must be included. The production rate is set for optimum yield and minimum "downtime;" that is, non-productive time. Process controls are usually centralized and may be highly automated. Although in-line mixers reduce first cost by eliminating tanks and simplifying the pump-and-piping system, the more sophisticated controls they require are expensive.

The ingredients of the emulsion must be pumpable and reasonably homogenous to be processed in a continuous manner.

The flow rates must be known, and the retention times must be established accurately to determine the design of equipment. Scaling up from pilot plant data is more difficult in a continuous process than in a batch process. Although the rate of production can be altered by altering the rate of flow, the residence (or blending, cooling, etc.) time is also altered. Altering the rate of flow of a continuous unit may result in a different product. It should be noted that in extended continuous operation, impurities may build up under certain conditions, so that periodic shutdowns are unavoidable.

Advantages and Disadvantages of the Continuous Method

Continuous flow is worth consideration:[15]

1. If the reaction does not need to go to completion.
2. If volumes are large or time cycles are short.
3. If the materials are of low intrinsic value.
4. If countercurrent operation is feasible.
5. If a considerable excess of one reactant is allowable.
6. If precipitations, solutions, or extractions are involved.
7. If 24-hour operation is planned.
8. If much laboratory work is called for to determine end points of each batch.
9. If, for some reason, a considerable fraction of the time cycle of a batch must be taken up in filling and emptying the tanks.

Continuous flow is of doubtful benefit:[6]

1. If volumes are small, or time cycles are long, or both.
2. If operation is intermittent, as in plants running only 8 hours a day.
3. If complex cycles of temperature or pressure must be followed.
4. If undesirable bacterial action can take place under the operating conditions.
5. If a slow reaction must be go to completion.

Semi-continuous Process

An intermediate type of arrangement (*semi-continuous*) employs batch-type vessels and heat-interchanging equipment, operated on preset cycles of feeding and withdrawal. When low-viscosity ingredients are involved, a dissolving vessel with agitator may be required for proper timing of the operation. Retention time and process parameters can be varied within fairly wide limits. Process control is less sensitive than for a completely continuous operation.

TYPES OF EMULSIFICATION DEVICES

To produce a commercially satisfactory emulsion, adequate dispersion must be achieved at reasonable cost. Both first cost and operating cost must be considered. A product of the desired viscosity and stability should be obtained for the least total cost. A satisfactory product should be produced in the shortest time with the least power and heat in the cheapest possible equipment. Intelligent design and proper selection of construction materials will minimize requirements and costs of maintenance.

Stirrers or agitators provide, in a given batch, a hydrodynamic condition favorable to improved interchange of materials and heat. The homogeneous distribution and intimate mixing of ingredients in emulsification involve a reduction in particle size. In turn, reduction in size requires an application of shear. The pumping action and the reduction in size can be supplied by the same element (in heavy-duty equipment such as colloid mills, homogenizers, and vibratory devices) or by separate elements (as in agitators of various types).

Flow And Shear

An impeller is primarily a pumping element; the power input is proportional to the product of flow and the resistance offered by the liquid. This hydraulic shear (resistance to flow) depends upon the viscosity of the liquid and the restrictions to flowing caused by the passages through the mixer. Either flow or shear may be the major factor to be considered in an emulsion process. Therefore, the power to produce flow is proportional to the product of the rotational speed times the cube of the diameter of the impeller. The power to overcome shear is proportional to the product of the square of the rotational speed times the square of the diameter of the impeller. Hence, the over-all power requirement becomes proportional to the cube of the speed of rotation and the fifth power of the diameter of the impeller. For a given input of power, an increase in shear would require an impeller of reduced diameter and higher speed; for increased flow a larger impeller running at less speed would be indicated.

Power Requirements

Model experiments supply the specific output of power for the given design, speed of agitator, and size of vessel. Measurement of the time requirements and power inputs for various speeds supplies data on optimum speed of tip and size of batch. The speed of the shaft can be derived from that of the tip. With the power input known, the loss of power in the packing box can be estimated. This, in turn, yields the total load on the drive of the agitator. The torsional momentum and size of the shaft are a function of the total load on the drive and the material of the shaft. A rational procedure for determining power requirements of the propeller and turbine-type of agitators is presented by Parker using suitable nomographs.

As a rough approximation, Table 7:1 slows power requirements based on a processing time of approximately hour a batch (or gal/hr in a continuous process).

TABLE 7:1

APPROXIMATE POWER REQUIREMENTS (horsepower)

Batch Volume, gallons	*Propellers*	*Turbine Mixers*	*Homogenizers*	*Colloid Mills*
2		0.25		0.5
15	0.1			2
25	0.25		1.5	5
50		1		
200	0.5	5		10
750	1	10	25	
1000		15	30	
1500	2	20		
2000	3	25		80

MIXING TANKS

Generally, for the making of emulsions, vertical cylindrical tanks with dished or spherical heads are preferred for accessibility, maintenance, and convenience in design and fabrication. Open tanks serve commonly as premixers and for simple blending or emulsification of low-viscosity materials. An open tank should be reinforced at the rim, or some other means of support provided, for clamp-on agitators. Closed vessels are used for operation at other than atmospheric pressure or to control the emission of flammable or otherwise hazardous materials.

Equipment and tank covers are often supplied with sight-glasses and may also have a lighting fixture to aid the observation of the process. Mixing tanks are usually supported on pipe legs screwed into sockets welded to the bottom of the tank. The legs should be adjustable to accommodate the slope of the floor. Jacketed vessels consist of an inner tank and an outer shell to contain the heating or cooling material that is circulated through this annular space. The shell is usually bolted or welded to the inner tank, and the whole unit is supported by a steel structure. The jacket may be continuous or consist of a series of channels or tubes.

Although baffles in the mixing vessel are not usually required for products of high (greater than 50,000 centipoise) viscosity, four vertical internal baffles are usually supplied to provide proper agitation of most other emulsions.

It goes without saying that all thermometer wells, baffles, and heating coils should be fastened securely in place in order not to interfere with the operation of the impeller.* All equipment for any process should be smooth inside (at least) for ease of cleaning.

The capacity of agitator vessels should be sufficient to hold the charge for one run with 10 to 30 percent excess. The overage allowance will depend on the characteristics of the emulsion. Some vessels must be designed to ensure good mixing for both the initial volume and the volume of an eventual batch if the ingredients are added stepwise.

Retention Time

The *retention time* is the time allowed for the completion of the mixing and emulsifying procedure. Either the capacity of the container must be sufficient for the desired retention time, or the retention time must be computed from the capacity of available equipment. The volume of the charge in a kettle should seldom exceed 80% of its rated capacity. If V is the volume of the charge in cubic feet, F is the feed

* This statement seems obvious, but many sets of baffles have been removed from kettles because the clearances were not sufficient.

rate in lb/min, v is the specified volume of the feed in cu ft/lb, T is the hypothetical or desired retention time in minutes, and s is the space velocity in cu ft/min, then the retention time is

$$T = \frac{V}{v} F$$

and the space velocity is

$$s = v \frac{F}{V}$$

Mixing Apparatus

The agitator should be designed for optimum mass transfer, another way of saying *mixing efficiency*. The time of agitation is merely a function of time in a batch process, but in a continuous process it is dependent upon the design and volume of the mixing chamber. The size and shape of the tank and the viscosity of the emulsion influence the size, type, and location of the impeller. Inadequate correlation of the size and shape of impeller and kettle may cause mixing problems called

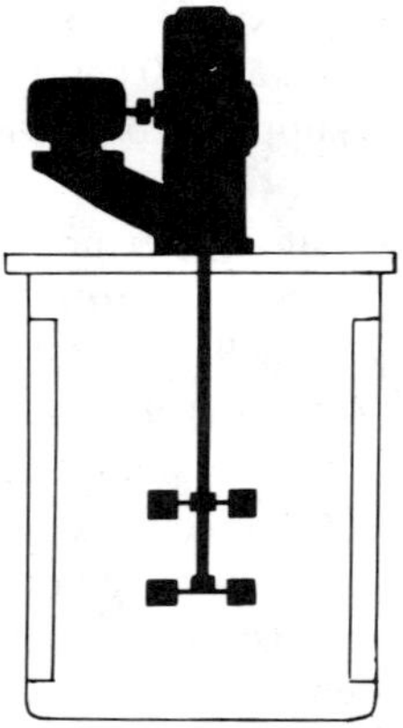

Fig. 7:3. CROSS SECTION OF OPEN TANK WITH TWIN IMPELLER-MIXERS AND INTERNAL BAFFLING.

"proximity effects." Proximity effects are characterized by excessive power consumption and/or a waste of processing capacity.

The ratio of the diameter of the tank to the diameter of the impeller (T/I) is an important consideration in design. This ratio, combined with the speed of the impeller, determines the shear-to-flow ratio for a given system. Tall and slender tanks are suitable for low-viscosity emulsions and high-speed impellers. If the height of a tank exceeds 1.5 times its the diameter it requires multiple impellers (see Fig. 7:3). More viscous materials, mixed in tanks of which the diameter exceeds the height require lower-speed impellers. The effect of the shape and the proportions of the enclosure on the dispersing action is proportional to the viscosity of the emulsion. Stator blades may be required to control the shearing action of the impeller. Propeller or turbine impellers are commonly used for emulsification. The following ratios in diameters of tank-to-impeller provide satisfactory first-order approximations for such impellers.

Viscosity	*Ratio in Diameters Tank:Impeller*
0 to 15,000 centipoise	2.5 to 3 (exceptionally to 6)
15,000 to 25,000 centipoise	2.0 to 2.5
above 25,000 centipoise	1.5 to 2.0

Location of Agitator

The location of the impeller in relation to the inside of the kettle is also influenced by the diameters of tank and impeller. Limitations have been established for the desirable *submergence* (distance between impeller and surface of liquid) and the *bottom clearance* (distance between impeller and bottom of mixing vessel). For propeller agitators, the normal submergence is at least 1.5 times the diameter of the propeller. For the same agitator, a bottom clearance of one or two diameters is common. A submergence of two or three diameters is used for impeller agitators. The bottom clearance for impellers is 1.25 diameters unless the depth of the liquid is less than three diameters, in which case the bottom clearance is one-third the depth of the fluid is used. Parker has outlined a procedure for the determining the proper relationships between characteristics of impeller and container.[7]

Shafts

Mixer shafts must be rigid to exclude vibration. The operating speed must stay below the threshold of critical vibration. Shafts must must also run true to maintain the flexibility of the packing. Propellers pass quickly through the first critical vibration frequency and are usually operated between the first and second critical vibration frequencies. The material of a stirrer must be substantial enough to withstand the most viscous component of the emulsion and the "thick phase," if one is expected.

If the shaft of the propeller is long and inadequately supported, the critical conditions can be attenuated through the use of stabilizing rings attached to the propellers. These stabilizing rings are typically one-quarter the width of the outer diameter of the propeller and are attached concentrically with the shaft at centerline. The weight of the shaft in large mixers can be effectively reduced by the use of heavy-wall tubing rather than a solid shaft. The use of tubing raises the critical speed of the shaft because the mass is reduced in greater proportion than the strength. One degree of torsional deflection on twenty diameters of shafting is considered admissible in the design of a shaft. Both sleeve and flange couplings are used on agitator shafts, but pins or keys should be provided to prevent the shaft from dropping into the mixing tank. Enameled and rubber-coated agitators are often connected to the shaft by an inner bar threaded on the inside. In other cases, the shaft is fitted with a sleeve attached to the impeller.

The shaft opening in an open tank may be protected by an umbrella collar mounted on the shaft. Wiper contact seals with low-pressure differentials suffice for closed tanks. Packing boxes fitted with proper lubrication are adequate to about 150 *psi*. Stuffing boxes should permit repacking without dismantling the drive. The packing must not score the shaft and should not be used as a shaft guide.

Gears and Bearings

Helical-type gearing is preferred for many operations. The ratios of the gears should be so selected that standard motors can be employed. As gears, bearings, and shaft seals contribute to loss of power they thus should be designed for minimum loss and should be lubricated properly. Hydraulic drives are gaining acceptance, particularly for low-speed, high-torque applications. Mixer bearings must be designed to withstand both axial and radial forces; also normal shock loads and expected contamination. Vertical loadings are usually supported by anti-friction bearings; radial loads, by sleeve-type bearings. Heavy-duty roller bearings are specified for extremely high loadings, as in homogenizers. Bearings should not be submerged in the batch, and this is rarely necessary; but if it is unavoidable, such bearings must be easy to clean. If the chemical nature of the emulsion permits, plastic bearings can be used in connection with hard shaft sleeves. Under no circumstances should the bearings form a galvanic couple with the other components of the vessel.

Source of Power

Both horizontal and vertical electric motors are used to drive pumps and agitators in the production of emulsions. Explosion-proof designs are necessary if flammable solvents are used in the area of operation. Protection against overloads is assured by the use of shear pins or equivalent means.

Switches, starters, valves, and other control elements are advantageously combined on a welded steel frame close to the kettle.

Heating and Cooling

Most of the power supplied to high-shear emulsifying equipment is dissipated in the form of frictional heat in both the machinery and the emulsion. The amount of frictional heat formed during high-shear emulsification is proportional to the viscosity of the emulsion at the time of mixing. When chemical reactions occur, the energy content in the reaction product will usually be different from that of the reactants. The emulsion that forms with a chemical reaction taking place will usually gain heat because such reactions are exothermic. The resultant increase in temperature will not only speed up the reaction, but will also tend to destroy or weaken the emulsion. Therefore, external cooling or the slow addition of one of the reactants is necessary to control the temperature of an exothermic reaction.

On the other hand, the addition of heat may be necessary to melt one of the components of the emulsion or decrease its viscosity. An increase in temperature of one degree Centigrade causes a decrease in viscosity of ten percent in certain organic materials. Temperature control will often have to be maintained for some period of time. In batch processing a specific temperature pattern may have to be repeated according to a fixed schedule. Automatic recording and controlling instruments are often used to maintain fixed temperature patterns.

The properties of many emulsions are affected not only by the temperatures employed during the manufacture but also by the rate of heating and/or cooling. Greases and many other emulsions can change their flow properties or stability if they are cooled too fast. The rate of heat transfer from the heating or cooling medium to the emulsion is a function of the differences in temperature between them, the heat conductance of the wall of the kettle, the surface area available for the heat transfer, and the flow velocity of each fluid across the heat transfer zone.

Sources of Heat

In the manufacture of emulsions, steam is probably the most common heating medium. It is delivered to the upper portion of a jacketed kettle, and condensate is drained from the lowest point in the jacket.

The flow of cooling water through the jacket must be reversed.

An organic compound, such as Dowtherm®* is also used for heat-transfer. The Dowtherm is brought to the proper temperature in an external heat-exchanger. Temperature can be accurately controlled through use of such a system. When the heat that can be supplied through the jacket of the kettle is insufficient, or jacketed kettles are not economical for the specific process involved, internal heating elements may be used. For maximum efficiency, such heating elements are are shaped and arranged in the kettle to act as baffles. The heat-transfer elements in the kettle may be shaped as a vertical curtain, or be concentric with the wall of the kettle.

Depending on the surface area needed for heat transfer, the concentric coils may have the shape of a single-or multiple-layered cylinder. Internal heaters and baffles are difficult to clean, so they are often positioned a short distance from the wall of the kettle to reduce the accumulation of deposits. Removable baffles and internal heaters are also used to permit easy cleaning. Flexible piping with "quick disconnects" permits a high degree of flexibility in the plant layout at relatively minor cost.

The total heat requirements are computed from the specific heat of the emulsion, the mass of the emulsion, and the temperature to be maintained or achieved in the emulsion. The heat-transfer rate is increased by thinner walls between the heat-transfer fluid (steam, water, Dowtherm) and the emulsion, and by more turbulent conditions in the emulsion and in the heat-transfer fluid. Emulsion polymerization has found favor because the heat of reaction is quickly dissipated from the reactants into the mass of the continuous phase.

The design of the heat-transfer surface requires special care because many emulsions are sensitive to loss of stability if subjected to wide fluctuations in temperature. Localized boiling, for instance, will destroy many emulsions, or the part of the emulsion that is subjected to the boiling. If an external heat-exchanger is used, the hot and cold fluids should enter and leave the heat exchanger so that the cooled water is leaving. The cooled emulsion then leaves the exchanger at the end where the cold water is entering. Maximum efficiency from the heat-exchanger and minimum thermal shock to the emulsion result if this procedure is followed. In a batch process, the agitation in the kettle helps to increase the rate of heat transfer and minimize localized heat-

* The Dow Chemical Company, Midland, Michigan.

ing or cooling in the emulsion. Continued agitation during heating or cooling is usually beneficial. Special equipment for heat-transfer is more efficient than most kettles, but it is expensive and unnecessary for most small-volume production of emulsions.

Insulation and Heating of Pipes and Vessels

Under average conditions the base walls of process equipment may radiate and lose five to ten percent of the process heat. Proper thermal insulation around the exterior of pipes and vessels will reduce these losses to around one percent and minimize the thermal burns to production personnel. If process equipment handles hot emulsions, the saving that result from the investment in insulation are often considerable. Hair felt, cork, and certain types of glass wool may attack aluminum equipment, so the choice of insulating material should be a judicious one. The best over-all type of insulation is probably the composites of aluminum foil. They are clean, easily removed and reapplied, and insensitive to vibration, and they are good thermal insulators. Insulations of the panel type, or sheet-metal boxes of the proper curvature to fit the equipment and filled with an insulator are also used.

If hot liquids are to be transported through piping, the pipelines should be jacketed (insulated) to prevent heat loss from the contents of the pipe.

If the material in a pipe is in danger of freezing, small steam lines of steel or copper would be laid next to the pipe. Tubing and pipeline are then insulated as if they were one pipe. Electrical heating tapes are also used to prevent solidification of material in pipelines due to freezing. Pipelines containing either hot or cold materials should be laid with expansion loops or other means to compensate for linear thermal expansion and contraction.

Pressure and Vacuum

Most emulsions are produced at atmospheric pressure because of lower costs of equipment, maintenance, safety, and simplicity of design. Operations at *reduced pressure* may be advantageous for one or more of the following reasons:

1) The relative differences in volatility between substances to be distilled are widened.

2) Excess water or other volatile ingredients can be removed by distillation with relatively little heat, and thus minimize thermal decomposition of the emulsion or its ingredients.

3) The boiling point of the mass can be reduced to allow use of

a relatively cool heating medium.

4) Reaction rates during emulsion polymerization can be controlled.

On the other hand, *increased pressure* may be desired:

1) To maintain temperatures above the boiling point of the emulsion or its ingredients; 2) To control rates of reaction that are pressure-sensitive either directly or indirectly; 3) To raise the operating temperature of the emulsion for the sake of optimum heat balance in the over-all system.

Process equipment must be designed for the specific conditions to be encountered. Thrust bearings on agitators are usually designed only for vacuum or for pressure. The use in a vacuum process of an agitator that was designed for pressure operation is normally "one time only." Sampling a product, adding anything to an emulsion, or withdrawing it while under pressure or vacuum, increases operating costs. This factor, plus the higher initial capital investment, must be balanced against the advantages to be gained from either a pressure or a vacuum process.

PUMPS

Pumps are selected according to the range of capacity expected, properties of the feed, and conditions of suction and discharge as well as the range of temperature. Higher temperatures require cooled bearings and sealing devices; mechanical seals are used above 90°C. All pumps must be easy to inspect and clean.

1) For *centrifugal pumps*, design depends on type and speed of the impeller, the number of stages, and the viscosity, clogging tendency, and plasticity of the emulsion. Centrifugal pumps produce an essentially constant pressure when operating at constant speed. The volume delivered depends on the restriction at the outlet. Hence, the delivery is controlled by throttling, or by regulating the speed. Viscous liquids require a positive suction head (flooded inlet).

2) *Positive displacement pumps* are self-priming. The displacement of *reciprocating pumps* follows from the diameter, stroke, and speed of the piston. The displacement multiplied by the volumetric efficiency is the capacity of the pump. As viscosity increases, the volumetric efficiency decreases.

The flow from a pump can be adjusted by means of a variable-speed transmission, by changing the length of stroke, or by returning any excess delivery to the suction side of a relief valve.

3) *Diaphragm pumps* may be considered where the materials handled contain gritty or lumpy solids, or are viscous, corrosive, or unstable.

4) *Rotary pumps* provide a steady flow, but are more sensitive to abrasive emulsions, because their efficiency depends on the clearances between rotors and housing. Also the material being pumped should have lubricant properties. *Lobe pumps* with timing gears avoid metal-to-metal contact, and thus can handle abrasive and highly viscous slurries.

5) *Metering pumps* are reciprocating units characterized by a variable crank or throw arm, to which linkage operating the plunger is connected. The body of the pump is end-packed and equipped with one or more sets of suction and discharge valves. Free-acting check valves permit a metering accuracy of 2—3 per cent. Mechanically controlled valves give accuracies of 0.1 to 0.2 per cent. If suspensions are present, however, or if the viscosities vary, accuracy is difficult to maintain.

The body of the pump must be designed to permit a choice of diameter for the plunger and economical fabrication from various materials of construction. Also, the materials of the mechanical components must be chemically and physically compatible. Packing boxes must be of liberal dimensions, and the packing should be selected to withstand chemical deterioration in service.

Delivery from the pump can be checked by mounting the supply vessel on a scale, with a flexible connection between it and the pump.

Proportioning systems basically consist of two or more metering pumps adjusted to handle variable quantities of fluids according to formulation proportions. The individual streams are then controlled by devices for measuring flow, for timing, or for programming, depending on the established schedule.

CONVEYING OF SOLIDS

Solids ingredients in larger quantities can be moved by *pneumatic handling*. Conveying lines are flexible, and different materials can pass through the same system without appreciable contamination.

Negative-pressure systems generally convey from several pickup points to one delivery point. A vacuum pump (centrifugal fan or rotary blower), separator (cyclone collector) and dust filter are required.

Positive-pressure systems move solids from one supply point to several discharges. In addition to an air compressor (blower), air locks are needed.

For dry ingredients, components of a spouting system, such as branch-offs, line diverters and sight tubes, are available in light-weight construction, and with cleanable and renewable sighting sections made of plastics or tempered glass. They can be equipped for remote operation by power cylinders or solenoids. Rotary air locks are built for drop-through or blow-through service, and they are usually rated by the volume displaced in one turn. They have round inlets on materials, outboard bearings, and sometimes air-purged glands.

As most emulsification processes are batch operations, the ingredients are usually weighed into the kettle. Most formulations, therefore, are given in terms of weight percent. For added accuracy, many production plants use a "fill to the line" procedure; that is, the operator fills a previously calibrated container to a certain line and empties the entire contents into the mixing kettle. The latter may be mounted on a scale set into the floor of the prodcuction building and the weights determined from this setup. Pre-weighing into a container that is later emptied into the batch is much more accurate because it guards against adding an excess of any component. More sophisticated techniques of measuring the flow of fluid are available for the larger installations and continuous processes.

MATERIALS OF CONSTRUCTION

Materials of construction for mixing vessels, lines, pumps, gauges, etc., are selected according to:

1) Strength and rigidity
2) Resistance to deterioration in service
3) Machineability and weldability
4) Recovery value

High-carbon and high-chromium irons or chilled white iron may serve where abrasive action occurs. To resist the continuous impact of particles, austentic steels are employed. Certain magnesium steels harden under impact. As a rule, materials suitable for one condition can not be assumed to be useful when applied to another condition.

Ferritic-type stainless steels must be heat-treated whenever any welding or other thermal shock is applied to a piece of equipment. Welding of only a small part to a large unit requires heat treatment of the whole unit. Stabilized stainless steel can be welded without extensive heat-treatment.

The first cost of tantalum equipment is high, and its scrap value is low. Conversely, the recovery value of silver is 80 to 90%. The demands of most emulsions are moderate on materials of construction.

METALS

Many metals and alloys are available for the construction of an emulsification plant. The best metal for each piece of equipment is the one that provides most satisfactory performance at least cost. The resistance of metals and alloys to their environment often limits their serviceability for emulsification.

Aluminum equipment is good for distilled water, ammonia compounds, benzene, benzoic acid, lactic acid, salicylic acid and other chemicals. It should not be used for halogen acids or alkalis. For lines for pine oil and resin, aluminum is preferred because of its colorless salts.

Carbon steel serves for tar products, ethers, oleic acid, oxalic acid, and others. Copper-bearing carbon steels are used for heating and cooling jackets.

Cast iron is good for alcohols, fatty acids, acetic acid, and similar materials. Core plugs in cast parts may cause eletrolytic corrosion. *Silicon iron* (with 12% or more of Si) is used for halogens, oils, fats, butyric acid, and certain other chemicals.

Alloy steels depend for their chemical resistivity on their composition, mode of manufacture, surface treatment, and other factors. Stainless steels may be of *ferritic* (4 to 23% Cr) or *austenitic* type. They serve well under oxidizing conditions, but "stainless" properties may be nonexistent in the presence of reducing agents. An alloy steel with 18% of chromium withstands acetic acid, lactic acid, ammonia compounds, etc.

Stainless 304 and 308 are good for acetic acid and 100% acetic acid vapor; types 316 and 317 are relatively resistant to dilute vapors; type 316 also series for formic acid and its solutions.

Stainless 304 is used for soaps, edible fats, emulsions of oil and of wax. Type 316 serves for vegetable and animal oils, and some concentrations of sulfuric acid, temperature permitting.

The Hastelloys have high over-all chemical resistance but they are expensive. Hastelloy A is satisfactory for hydrochloric acid, but

Hastelloy B is affected by it in the presence of oxygen. Hastelloy C resists hydrochloric acid of all concentrations.

Stellite-coated and solid stellite components are found in general-use colloid mills.

Lead must be desilvered to resist corrosion. Its low tensile strength and tendency to creep prevent use at high temperatures. In combination with antimony and tellurium, it can be used will alkalies, and with hydrofluoric, tannic, phosphoric, boric, and other acids. "Crawl-proof" lead is antomony surfaced with lead. Iron parts are often pickled to remove scale and dust, then dipped in antimony, and finally in lead.

Copper, *brass* and *bronze* are used with alcohols, organic solvents, and also acetic acid, if air is excluded. Tubes are usually seamless; clamping nuts and such parts exposed to contact with process materials are made of Everdur or equivalent alloys.

Preferred cupro-nickel alloys include the 70-Cu/29.5-Ni/0.5-Fe type, and the 70-30 type. They are relatively immune to stress corrosion by moist ammonia gas. Nickel, Inconel, and Monel are unaffected by most organic solvents, oils, and alkalis. Halogen compounds, phosphoric and picric acid, and carbon tetrachloride should be avoided. Nickel is satisfactory with phenol.

Invar, which has negligible thermal expansion up to 200°C, is used in colloid mills equipped with heating or cooling jackets to maintain close milling tolerances.

Steel clad with a facing of nickel, Monel, or the like may prove economical, especially for vacuum equipment. The corrosion-resistant facing usually corresponds to 10 to 20% of the total thickness. The minimum practical thickness of clad steel is about 0.25 inch.

Silver can be used with acetic acid if no chlorine is present. It should not be used for halogens, in general.

Tantalum equipment is used in contact with hydrochloric acid, sulfuric acid, perchloric acid, hydrogen peroxide, iodine, bromine.

Tinned equipment is used for essential oils, sulfites, pharmaceuticals.

Non-Metals

Soda and quartz glass are used for small-scale equipment. Glass-lined or enameled iron and steel equipment is used in the vacuum processing of organic materials. Glass will not resist hydrofluoric acid, and may be attacked by sulfuric, boric, and phosphoric acids and strong alkalis. Special attention must be given to sealing and joining faces and to thermal and mechanical stresses in the use of glass-coated equipment.

Stators in colloid mills have been made of fused aluminum oxide. "Rockide" coating is often applied to rotors in colloid mills.

Porcelain and stoneware parts are available in different grades, with specific chemical and physical characteristics. Porcelain parts can be made with an accuracy of better than 5% for turned and molded shapes and 2% for extrusions. Close tolerances (0.01 mm) can be met by grinding. Porcelain can be made non-porous. It is attacked by hydrofluoric acid, but is relatively resistant to alkalis. Stoneware will not resist hydrofluoric acid, strong phosphoric acid and caustic alkalis. Both the body and the glaze of the ware may be made of inert materials.

Carbon, including graphite and Karbate, can serve for fatty acids, and for lactic acid to 50% concentration at temperatures to 170°C. Materials of a highly oxidizing character will attack Karbate. Sealing rings of carbon, applied against polished hard-metal bushings, are generally used in mechanical shaft seals.

Rubber-lined equipment may serve for temperatures to 70°C. It is satisfactory with hydrofluoric acid, but will not withstand nitric, nitrous, or mixed acids. Synthetic rubbers (Buna, Viton, Silicone, etc.) are used in mechanical seals of the O-ring type, within the range of normal temperatures.

INSTRUMENTATION IN PLANT

The preparation of most emulsions requires little instrumentation in the production unit. Many of the instrumental techniques that are discussed in Chapter 3 are used in production operations. The small production unit, especially, has little need for the extensive automatic controls that are available today. Many small producers, while confining their use of instrumentation to a common thermometer and a set of scales, consistently produce high-quality products. Most automatic equipment is designed to take the place of care and simple instruments. Needless to say, sophisticated automatic instrumentation will aid the production of a uniform product.

In industry, instrumentation is used to determine whether the product corresponds to the emulsion that was made in the laboratory or in the kettle yesterday, or to ascertain when to perform a specific operation. An industrial instrument must have adequate accuracy, precision, and range for the particular emulsion(s) produced. All operating instruments should be easy to clean, and resistant to corrosion by the materials to be tested. Instrumental measurements are used for quality control (or assurance) in both batch and continuous production. The most commonly measured parameters in the emulsion plant include: temperature, *p*H, water content (or percent of one phase or the other), pressure, and time. The particular instruments are discussed in Chapter 1.

Industrial thermometers may differ somewhat from those commonly used in the laboratory or home. The thermo-couple is probably the most common device for measuring temperature after the common liquid-in-glass thermometer. The thermo-couple is unbreakable, within the limits of normal use, and is easy to use. Resistance thermometers and metallic expansion thermometers are used less widely. As most emulsions are produced in batches at temperatures between 25 and 90 degrees Centigrade, the common mercury-in-glass thermometer is adequate for most operations.

Controllers

For best results, the controlling elements must be designed to match the particular process and system to be controlled. The diversity of control variables requires selection from a variety of systems and components.

The basic techniques of control are: 1) On-off control, and 2) Continuous control. In the former, the final control element is moved from one of two fixed positions to the other, as from "on" to "off", or from "maximum" to "minimum", or vice versa. Where the energy-storage capacity of the equipment is considerable, this kind of control is usually adequate. With continuous control, the final control element may assume any position within the operating range of the controller.

In continuous control, the level of the liquid in a vessel may be controlled by adjusting the inlet valve via a lever directly connected to a float. Any variation in the position of the float then changes the manipulated variable, and the two are inter-related in a definite way within limits given by the "proportional band." Maintaining equilibrium between influx and discharge requires a definite set point. The proportional band can be adjusted by modifying the leverage ratio. This is called *proportional control.*

The pressure inside a vessel is often controlled by a pressure gauge acting upon a potentiometer. The potentiometer varies the voltage fed to a servomotor that operates the inlet valve. The magnitude of the voltage controls the speed of the servomotor. The value of the correcting impulse thus depends on the deviation and the duration of it. Such a floating control may be converted into proportional control by introducing a constant feedback.* Then, the corrective active is always cognizant of conditions at the control input.

The following terminology for process control is included for reference:

An *automatic controller* is a device that measures the value of a variable quantity or condition and operates to correct or limit deviation of the measured value from a selected reference.

The *control elements* comprise that portion of the controlling system which includes action relating the manipulated variable to the actuating signal.

The *controlled system* is the process, or machine, in which a particular quantity or condition is to be controlled.

The *controlling means* are the elements of an automatic controller that are involved in producing a corrective action.

Deviation is the difference between the value of the controlled variable and the set point.

Feedback elements comprise that portion of the controller that establishes the relationship between the primary feedback and the controlled variable.

The *final control element* is the portion of the controlling means that directly determines the value of the manipulated variable.

The *input elements* comprise the portion of the controller that provides the reference input in response to the set point.

The *manipulated variable* is the quantity or condition that is varied by the automatic controller so as to affect the value of the controlled variable.

The *reference input* is a signal established as a standard of comparison for a feedback control system by virtue of its relation to the command.

A *servomechanism* is a feedback control system in which the controlled variable is mechanical position.

The *set point* is the selected reference value of the controlled variable which it is desired to maintain.

The *primary feedback* is a signal that is a function of the controlled variable and which is compared with the reference input to obtain the actuating signal.

* Constant feedback means feedback invariable with time.

PACKAGING MACHINES

Packaging machinery consists of synchronized individual units that fill cans, bottles, or collapsible tubes with the finished emulsion. Such units may be manually operated or partly or fully automatic.

Viscous liquids are usually metered by piston pumps. Weighing-filling units have a scale beam that closes the filling spout when the preset weight has been dispensed. Collapsible tubes are filled from the bottom, or crimp end, to prevent inclusion of air bubbles. Preheating of an emulsion is occasionally necessary to provide more desirable flow characteristics. Highly viscous materials may have to be supplied to metering pumps by feeder agitators. Separate units are necessary for closing or sealing and wrapping the filled containers.

Pharmaceutical preparations, especially those requiring a high degree of purity, are packaged in "clean rooms" under sterile conditions. If the emulsion or its ingredients have a tendency to react with common construction materials, the packaging equipment should be of glass, stainless steels, or Teflon. The equipment must be so designed that it can be readily disassembled for cleaning and inspection.

A complete guide to packaging equipment can be found in the annual Modern Packaging Encyclopedia (Published by McGraw-Hill, Inc., New York).

References

1. PERRS, J. H., editor, *Chemical Engineers' Handbook*, (3rd Ed.); McGraw-Hill Book Company, Inc., New York, 1950.
2. MCCABE, W. L., and SMITH, J. C., *Unit Operations of Chemical Engineering;* McGraw-Hill Book Company, Inc., New York, 1956.
3. PETERS, M. S., *Plant Design and Economics for Chemical Engineers;* McGraw-Hill Book Company, Inc. New York, 1958.
4. ARIES, R. S., and NEWTON, R. D., *Chemical Engineering Cost Estimation;* McGraw-Hill Book Company, Inc., New York, 1955.
5. PERRY, *op. cit.* (1) p. 1230.
6. KENDALL, *Chemical Engineering*, February 20, 1961.
7. PARKER, *Chemical Engineering*, June 8, 1964, pp. 177-82.

chapter 8

SELECTED TOPICS

I. FORMULATION OF EMULSIONS—HLB

The end result of all emulsion technology is the formulation and sale of a complete, satisfactory emulsion. The following steps are usually taken in the development of the emulsion in the laboratory.

Type of Product

To determine what product you want to make is most important, but unfortunately this obvious step is often not given enough consideration. Knowing exactly what type of product you wish to make will profoundly affect your result.

Purpose of Product

It is also most important to determine what you want the product to do. The various requirements for the success of the product must be enumerated. Frequently these considerations will be subjective as well as objective; for example, in the formulation of a shampoo. The appearance and foaming ability of a shampoo will contribute as much to its sale as the effectiveness of the product to clean the hair and leave it manageable. Along with the list of subjective and objective factors, you must include a corresponding list of the methods of measurement to be used in determining the characteristics of the product. Allied with the system of measurement, include the standards that the product should meet. A shampoo, for instance, must clean the hair, but not too much. You must establish both the qualitative and the quantitative goals that are to be the objectives in the development of the product.

Choice of Ingredients

Select the Major Ingredients. Having determined what you are going to make and how you will know when you have made it, turn your attention to the selection of ingredients. Each ingredient should perform a useful function in the formulation. Your shampoo should contain lanolin, dried egg, silicone fluid, or other ingredients to provide manageability. You will use water as a carrier. Thus your emulsifier must form an oil-in-water emulsion. You may want to buffer the system at the pH of the skin. The ingredients will have to be safe for use in a cosmetic of this type and should be commercially available. The amount of each of the major ingredients must be tentatively established.

Choice of Surfactants—HLB

Determine the Surfactants to Use. The type of surfactant and the amount of it can be determined through use of the formulations in Volume 2; or surfactants can also be selected from the thousands available by the HLB system (Hydrophile-Lipophile Balance), developed by Atlas Chemical Industries to aid in the selection of surfactant(s).

The HLB is the weight percent of the portion of the hydrophilic surfactant divided by five. The HLB of an oil-soluble surfactant will tend to be low (0-10); the HLB of a water-soluble material will be high (10-20).

The oils that you wish to emulsify can also be classified on this scale by the HLB required by a surfactant to emulsify them. The HLB can be determined for either ingredients or surfactants by comparing them (in a simple oil-water-surfactant emulsion) with materials of a known HLB. Atlas Chemical Industries has prepared booklets that describe the HLB system and list the HLB for many materials.[1]

The HLB for surfactants can also be determined by Gas-Liquid Chromatography.[2] The HLB can be calculated for most polyol fatty acid esters as follows:

$$\text{HLB}=20\left[1-\left(\frac{S}{A}\right)\right]$$

where S=saponification number of ester

A=is the acid number of the recovered acid.

As the HLB is related to the weight percentage of the hydrophilic material, the HLB can be estimated from the molecular structure. A surfactant that is 84% hydrophilic would have an HLB of 16.8; for example:

Assume a material having a molecular weight as follows:

Sorbitan	164
Lauric acid	200
Ethylene oxide, 20 moles	880

Water of esterification	$-$ 18
	1226

Of this, the hydrophilic portion would have a molecular weight of

Sorbitan	164
Ethylene oxide, 20 moles	880
	1044

The HLB number of this material would equal:

$$\left(\frac{1044}{1226}\right)\left(\frac{100}{5}\right)=17$$

Analytical determination of the HLB for this material yields a value of 16.7.

These latter two relations are applicable only for nonionic emulsifying agents and some nonionics do not conform to these rules.

Once the required HLB of the oils is determined, the HLB of the emulsifying agent(s) should equal this figure for best stability. If a blend of emulsifying agents (X,Y) is to be used, the HLB is assumed to be an additive quantity. Thus:

$$\% \times (\text{in the blend}) = \frac{100\ \ (\text{HLB req} - \text{HLB}_{Y})}{\text{HLB}_{X} - \text{HLB}_{Y}}$$

The HLB does not reveal how much of the emulsifying agent (s) to use (2-5% emulsifying agent 30% oil is a good starting point), what chemical type(s) to use (try several types or blends with the same HLB), or what class of emulsifying agent to use (blend nonionics with either cationics or anionics, do not blend anionics with cationics). The HLB does provide a systemized method of reducing the number of possible substances to be evaluated in the laboratory.

The art of emulsification is never more in evidence than in the selection of one or several emulsifying agents to produce an emulsion with the desired stability. As experience is a great help in determining these variables, the experience of a number of technologists is included in Volume 2. The formulations presented there provide a second means of narrowing the search for the emulsion system. Such basic formulations can be modified to correspond to the particular requirements of the specific emulsion under consideration.

Making the Emulsion

Through the use of a basic formulation from this book and/or the HLB system, a product—an emulsion—is now concevied on paper. The next step is to prepare the emulsion and analyze the resultant product. The procedure and technique will often be arbitrary, or based on the basic formulation. Now the art of emulsification expands. Let us say that the shampoo that you made was a two-phase dispersion that would not foam. You must now alter the procedure and/or formulation and remake the product. Scientific planning of experiments (e.g., Latin Square techniques) can be helpful, but often the formulating of an emulsion does not lend itself to such techniques. So the process of "prepare and evaluate" must continue until the formulation has been found that will accomplish the goals of the product.

Once acceptable emulsions are produced with the major constituents, the minor additives must be incorporated into the process. The shampoo, for instance, may also contain a foam booster, foam stabilizer, perfume, preservative, and color. After this trial-and-error procedure of *Edisonian Research* has been repeated, perhaps hundreds of times, the desired product should result.

Because an emlsion is often temperamental, the final product should be made several times by several different persons. If a uniform, acceptable series of products is now on hand, the formulator's job has been a success.

Importance of Shelf Life of Product

Now that the final emulsion has been prepared in some quantity, the samples should be placed in the containers in which the product will be sold. This salable product should then be placed "on the shelf" in the laboratory for extended aging. After the aging period has been completed, the emulsion is again checked against the requirements of the product for performance and stability.

Although the stability of the emulsion may be estimated on the basis of particle size, or data from centrifuging, there is no substitute for the actual aging of the product. Aging tests should be conducted at elevated and depressed temperatures as well as at laboratory temperatures. Freeze-thaw cycling is particularly destructive to an emulsion. Variations in temperature degrade the emulsion more than constant aging at extreme temperatures. For a product that is to be distributed at retail, the emulsion should remain serviceable for at least three freeze-thaw cycles, three to six months at 122°F (50°C), and one and one-half to two years at room temperature (74°F). Because the industrial customer is able to control the storage and will be willing (though not eager by any means) to take some care in the storage of the product, the standards may be somewhat relaxed depending upon the individual circumstances.

II. BIODEGRADABILITY

The formulator of any emulsion product that may find its way into the surface water of the country must be concerned with *biodegradability*. The United States lags behind both West Germany and the United Kingdom in concern for "cleaner" water. Not many formulators of emulsions have to be concerned with the problem of biodegradability at the present time; however, the subject of biodegradable surfactants is especially important to formulators of detergents, shampoos, bath oils, etc.

The urbanization and industrialization of the last few years have placed heavy demands upon water supply systems. The economical treatment of water to remove unwanted substances basically depends upon dilution with "clean" water after some physiochemical treatment. Historically, man and industry have left to Mother Nature the filtration, aeration, and dilution of the wastes of civilization. With increasing demands on the water supply, extensive post-and pre-treatment of industrial and municipal water supplies has evolved. The five methods of sewage treatment that are now employed are:

1. Activated sludge—One third of population:
2. Septic tank and percolation field—One third of population
3. Direct discharge—1/12 of population
4. Primary treatment only—One-fifth of population
5. Trickling filter plants (widely used in Europe)—One-eighth of population[1]

An activated sludge unit is essentially a continuous reactor in which the sewage, including the surfactant, is contacted with a concentration of bacteria in the form of sludge.[3]

The septic tank uses the bacteria in the raw sewage for decomposition of the wastes. Filtration through the soil removes solid or suspended matter.

Since World War II, the use of synthetic detergents has increased to the point at which the sewage treatment plants are unable to handle the materials. Many surface-active agents, such as the alkyl benzenesulfonates (ABS), are resistant to the usual means of sewage treatment. In the mid-1960's the surfactant industry responded to official and unofficial governmental pressure in Europe and in the United States to improve the biodegradability of detergent formulations.

The emulsion formulator that desires to manufacture biodegradable products, must solve the following problems:

1. Ascertain the biodegradable status of the major ingredients of his proposed emulsion;
2. Determine the level of biodegradability that is necessary and/or desirable. Simple, cheap laboratory procedures which adequately correlate with actual conditions of use are necessary.

At the present time both governmental and industrial personnel are attempting to establish universally applicable methods of testing and limits for biodegradability. The technical literature—such periodicals as *Soap and Chemical Specialties*—will keep the formulator in the United States up-to-date on the current state of the art. It is safe to say that the pressure on industry to do everything possible to reduce water pollution will increase in succeeding years.

The following tests to determine biodegradability have been developed: The percentage of surfactant degraded is determined indirectly by the volume and surface tension of the foam, directly by analysis with cobalt thiocyanate and methylene blue, as a function of time. A complete review of methods of determining biodegradability has been published by Schonborn.[4]

The major tests currently used include:

1. *Laboratory-scale Activated Sludge Unit* (batch or semicontinuous).[5,6]

   The surfactant is diluted in a synthetic or natural sewage and fed daily to an aerated bacterial system. The flocculated bacteria are separated from the carrier water one hour prior to the daily addition of material. The supernatant liquid is analyzed for surfactant by cobalt thiocyanate or methylene blue.

2. *River-Water Die-Away Test* (developed by Hammerton).[7]

   Twenty milligrams of surfactant are added to one liter of river water in an Erlenmeyer flask. The mixture is blended with a magnetic stirrer and allowed to stand undisturbed in the laboratory at ambient conditions. The system is sampled for analysis by cobalt thiocyanate or methylene blue, initially and periodically for a period of several weeks.

3. *Shake-Flask Test* (a screening test for unformulated anionics).[8]

   A flask containing 30 mg/liter of the surfactant in a synthetic growth medium is inoculated with cultured bacteria. The bacteria and surfactant mixture is agitated mechanically to mix the contents as well as to provide some aeration.

The mixture is periodically sampled to determine the degradation of the surfactant by analysis with either methylene blue or cobalt thiocyanate.

4. *Warburg Respirometer.*[9]

A flask fitted with a central well is immersed in a bath at constant temperature. The surfactant and bacteria that have been acclimated to the surfactant are placed in the flask. The center well is filled with potassium hydroxide to absorb the carbon dioxide generated by the degradation of the surfactant. The flasks are agitated to ensure thorough mixing and aid in the transfer of the gases. The flask system is completely closed so that the disappearance of oxygen can be calculated from the pressure drop of the system. By comparing the oxygen-consumption of the surfactant-bacteria system with the bacteria alone and the theoretical amount of oxygen needed for complete degradation, the amount of surfactant degraded as a function of time can be determined.

5. *Biochemical Oxygen Demand–Bottle Test.*[10]

A stoppered bottle is filled with oxygen-saturated, buffered water that has been seeded with sewage bacteria. The surfactant is added to the system and an oxygen measuring electrode is used to monitor changes in the dissolved oxygen content of the water. The degradation of the surfactant is compared to the degradation rate of an easily degraded material (e.g. glucose).

Generally, these tests call for reduction of 80 to 90% in the level of surfactant within a few hours or days. West Germany specifies that detergents must be 80% degraded in an activated sludge unit within three hours. The Soap and Detergent Association specifies that an unformulated anionic be over 90% degraded by the *Shake-Flask Test* in order to be considered biodegradable. A reduction in surfactant of less than 80% indicates that it is not biodegradable. Surfactants that are degraded 80-90% in the *Shake-Flask Test* must be degraded more than 90% in a Semi-continuous Activated Sludge Unit to be classified biodegradable.[11]

The literature offers many articles that provide good reviews of the subject of biodegradability.[11–13]

It should be noted that statements by the individual companies concerning the biodegradability of their products are based upon their own testing and interpretation of the results. Therefore the stated results for each surfactant are not necessarily comparable.

III. SOME REGULATIONS CONCERNING FOODS, DRUGS, AND COSMETICS

The marketing of foods, drugs, and cosmetics in the United States of America falls under the administrative control of the Food and Drug Administration, Department of Health, Education, and Welfare. The Food and Drug Administration (FDA) applies the Food Additives Amendment (1958) and related regulations. Knowledge of the pertinent regulations is necessary for anyone marketing products that are used as, or in contact with, foods, drugs, or cosmetics. The formulator of an emulsion, therefore, should maintain a knowledge of the ingredients that are permissible for use in each product with which he is concerned. The Food and Drug Administration maintains an advisory branch* to aid the formulator in determining the current status of a proposed product. Title 21, Part 121 of the *Compiled Federal Regulations* (21 CFR 121) concerns the use of food additives. The following definitions and explanations are important to an understanding of the terms that are commonly used in this field.

> A *food additive* is all substances......the intended use of which results or may reasonably be expected to result, directly or indirectly, in affecting the characteristics of food packed in the container. "Affecting the characteristics of food" does not include such physical effects as protecting contents of packages, preserving shape, and preventing moisture loss. If there is no migration of a packaging component from the package to the food, it does not become a component of the food and thus is not a food additive. A substance that does not become a component of food, but that is used, for example, in preparing an ingredient of the food to give a different flavor, texture, or other characteristic in the food, may be a food additive.[1]
>
> *Safe* means that there is convincing evidence which establishes with reasonable certainty that no harm will result from the intended use of the food additive.[2]

Regulations regarding both direct and indirect food additives are published in the *Federal Register*. Some substances were used safely as food additives for a period of years ("Generally Recognized As Safe") while others were used as food additives as a result of specific approval from the FDA or other government agency prior to the enactment of the Food Additives Amendment ("Prior Sanction").

---

* See p.138.

Therefore, the use of particular food additives may be permitted either as a *Direct Food Additive* or an *Indirect Food Additive;* or through a *Prior Sanction*, or because the material is *Generally Recognized As Safe.*

Because they were historically used with safety, some materials have been *Generally Recognized as Safe* (GRAS) for specific uses. For instance, the use of egg albumen is generally believed to be safe in moderate amounts for a food emulsion. This GRAS status is the result of agreement of men that are qualified by their scientific training and experience as experts in such matters. The Compiled Federal Regulations list some, but not all, materials that have GRAS status. The emulsifying agents that are so listed are shown in Table 8:1.[3]

TABLE 8:1

MATERIALS GENERALLY RECOGNIZED AS SAFE

Material	*Tolerance* %	*Limitations or Restrictions*
Cholic acid	0.1	Dried egg whites
Desoxycholic acid	0.1	Dried egg whites
Diacetyl tartaric acid esters of mono- and diglycerides of edible fats or oils, or edible fat-forming fatty acids		
Glycocholic acid	0.1	Dried egg whites
Mono- and diglycerides of edible fats or oils, or edible fat-forming fatty acids		
Monosodium phosphate derivatives of diglycerides of edible fats or oils, or edible fat-forming fatty acids		
Propylene glycol		
Ox bile extract	0.1	Dried egg whites
Taurocholic acid (or its sodium salt)	0.1	Dried egg whites

The manufacturer of the food item must determine the status of any ingredient. As a result, he may wish to ask the Commissioner of the FDA his opinion, or petition for approval of the additive as a matter of record.

Subpart C of the *Food Additives Section* deals with the use of food additives in the feed and drinking water of, or for the treatment of, food-producing animals. Many emulsions are used for the treatment of food-producing animals. Emulsions and surfactants are also used to wash fruits and vegetables and thus are discussed in this section. The use of Polysorbate 60[4] and 80[5], lignin sulfonates[6], and chemicals used in washing fruits and vegetables[7] are discussed in this section.

Direct Food Additives

Direct Food Additives are materials that are directly added to food.

The use of an emulsified defoamer in a carbonated beverage, or flavor in an ice cream would be considered such a direct food additive. Many surfactants as well as emulsions are used as direct food additives.

Subpart D, dealing with *Direct Food Additives*, includes the following materials that might be of interest to the producer of emulsion.

Polyoxyethylene (20) sorbitan tristearate,[8] Polysorbate 80,[9] Sodium lauryl sulfate,[10] Propylene glycol alginate,[11] Hydroxypropyl methylcellulose,[12] Hydroxylated lecithin,[13] Sorbitan monostearate,[14] Polysorbate 60,[15] Salts of fatty acids,[16] Chemicals used in washing fruits and vegetables,[17] Defoaming agents,[18] Calcium lignosulfonate,[19] Methyl ethyl cellulose,[20] Propylene glycol mono and diesters of fats and fatty acids,[21] Polyglycerol esters of fatty acids,[22] Dioctyl sodium sulfosuccinate.[23]

Indirect Food Additives

Subpart F of the same regulation deals with the *Indirect Food Additives*. These are materials that may be present in the food through contact with containers or equipment, and additives that otherwise affect food. Many emulsions that are used as an aid in processing a food are considered indirect food additives. An emulsion that is used as a mold-release in the manufacture of candy, or the emulsified coating of packaging materials or equipment, are examples of indirect food additives. The pertinent categories that may be of concern in the formulation of emulsions include:

Releasing agents,[24] Defoaming agents,[25] Emulsifying agents or surface-active agents,[26] Defoaming agents used in coatings,[27] Guar gum, modified.[28]

Before the Food Additives Amendment of 1958 was passed, the FDA and other Governmental agencies authorized the use of many substances as particular food additives. This authorization for the use of a material at a specified concentration or less in a specified product constitutes *Prior Sanction*. As materials with Prior Sanction are also usable under the GRAS status, there is some duplication. Again, referral to the FDA is the best advice for the formulator or manufacturer if doubt arises.

This discussion of the regulations assumes that the manufacturer will use good manufacturing procedures to compound a food, drug, or cosmetic from the permissiable additives used at or below the permissible concentration. One of the current points that is being emphasied by the FDA is this "good manufacturing procedure." The food, drug, or cosmetic, must be produced under sanitary conditions, with adequate quality control to assure that the product specified on the formulation sheet is equivalent to the product that is packaged. If the FDA finds that insanitary procedures have been used in the production of a food, drug, or cosmetic, they have indicated that they will take appropriate action.

The appropriate FDA regulations and interpretations of them are being constantly revised. The formulator or manufacturer of an emulsion that may fall within the purview of these regulations should obtain the most recent information to protect himself.

References

Part I

1. *The Atlas HLB System,* (2nd Ed.); Atlas Chemical Industries, Wilmington 99, Delaware.
2. Becher, P. and Birkmeier, R. L., Determination of Hydrophile-Lipophile Balance by Gas-Liquid Chromatography, *J. Am. Oil Chemists* Soc., 41:169(1964).

Part II

1. The Detergent Revolution, *Chemical Week*, 94 (22):111 (1964).
2. New Scuffle Over Syndets, *Chemical Week*, 97 (15):100 (1965).
3. Booman, K. A., *et al.*, Degradation Studies on Branched Chain EO Surfactants, *Soap Chem. Specialties*, January 1965.
4. Schonaorn, W., Seifen-Ole-Fette-Wachse, 88:870 (1962).
5. Borstlap, C., and Kooijman, P. L., *J. Am. Oil Chemists Soc.*, 40:78 (1963).
6. Booman, *op. cit.* (3).
7. Konecky, M. S. *et al.*, Paper presented to 36th Annual Meeting of the Water Pollution Control (Federal), October 6, 1963, Seattle, Wash.
8. Widening Split Over Syndet Softness, *Chemical Week* 96 (21):20 (1965).
9. Tergitol S., Company publication Form 41229A, p. 13; Union Carbide Corporation, Chemicals Division, 270 Park Ave., New York. N.Y. 10017.
10. *Ibid.*
11. Swisher, R. D., Chemistry of the biological degradation of boundry surface-active substances, *J. Am. Oil Chemists Soc*, 40:648 (1963).
12. Dicker, D. W. G., *Biodegradability of detergents, Rept. Progr. Appl. Chem.*, 48:334 (1963).

13. Review Series on Biodegradability, *J. Am. Oil Chemists Soc.*, 40 (11):(1963).

PART III

Write to Commissioner, Food and Drug Administration, Department of Health, Education, and Welfare, Washington, D. C. 20204.

1. 21 CFR 121.1(e); U S. Government Printing Office, Washington, D.C., March 28, 1959.
2. 21 CFR 121.1 (i).
3. 21 CFR 121.101 (d) (3).
4. 21 CFR 121.235
5. 21 CFR 121.236
6. 21 CFR 121.234
7. 21 CFR 121.1091
8. 21 CFR 121.1008
9. 21 CFR 121.1009
10. 21 CFR 121.1012
11. 21 CFR 121.1015
12. 21 CFR 121.1021
13. 21 CFR 121.1027
14. 21 CFR 121.1029
15. 21 CFR 121.1030
16. 21 CFR 121.1071
17. 21 CFR 121.1091
18. 21 CFR 121.1099
19. 21 CFR 121.1102
20. 21 CFR 121.1112
21. 21 CFR 121.1113
22. 21 CFR 121.1120
23. 21 CFR 121.1137
24. 21 CFR 121.2509
25. 21 CFR 121.2519
26. 21 CFR 121.2541
27. 21 CFR 121.2557
28. 21 CFR 121.2568

Selected Bibliography

P. BECHER, *Emulsions, Theory and Practice;* Reinhold Publishing Corp., New York, 10022, 1965.

An extremely well documented basic textbook concentrating on the theoretical aspects of emulsions.

F. A. BOVERY, I. M. KOLTHOFF, A. I. MEDALIA, E. J. MEEHAN, *Emulsion Polymerization;* Interscience Publishers, Inc., New York, 1955.

A thorough survey of the field of emulsion polymerization from a theoretical point of view.

J. F. DANIELLI, K. G. A. PANKHURST, A. C. RIDDIFORD (Ed.), *Recent Progress in Surface Science;* Academic Press, Inc., New York, 1964.

An excellent guide to the literature of the field of surface chemistry and the recent developments in the field, predominately from the theoretical point of view.

J. W. MCCUTCHEON, *Detergents and Emulsifiers*, 1965 (published annually); John W. McCutcheon, Inc., Morristown, N.J., 1965.

An invaluable, always up-to-date, list of surfactants, their physical form, and general area of use.

J. L. MOLLIET, B. COLLIE, W. BLACK, *Surface Activity*, (2nd Ed.), D. Van Nostrand Co., Inc., Princeton, N.J., 1961.

A thermodynamic and physicochemical analysis of surface and interfacial tensions and wetting that includes and excellent review of the industrial preparation of surfactants.

L. I. OSIPOW, *Surface Chemistry;* Reinhold Publishing Corp., New York, 1962.

A well documented survey of the theoretical and practical aspects of surface chemistry.

K. SHINODA, B. TAMAMUSHI, T. NAKAGAWA, T. ISEMURA, *Colloidal Surfactants;* Academic Press, New York, 1963.

An excellent theoretical approach to micelle theory and surface phenomena.

List Of Emulsifying Agents

Name	Supplier	Anionic					Nonionic				Cationic			Natural					
		Sulfates	Alkyl Sulfonic Acids and Their Salts	Carboxylic Acids and Their Salts	Alkylaryl Sulfonic Acids and Salts	Miscellaneous	Polyalkoxy Ethers	Alkoxy Esters	Alkanolamides	Miscellaneous	Amines and Amine Salts	Quaternary Amines and Their Salts	Miscellaneous	Amphoteric	Cellulose Derivatives	Water Soluble Gums	Lipids	Fluorocarbons	Silanes and Silicones
Abex	Alcolac Chemical Corp.					×													
Actiflo 68, 70	Central Soya Co.																×		
Acto	Humble Oil and Refining Co.				×														
Ahcol	I.C.I. Organics, Inc.	×																	
Ahcovel FP	I.C.I. Organics, Inc.							×											
Ahcowet ANS	I.C.I. Organics, Inc.				×														
Ahcowet RS	I.C.I. Organics, Inc.	×																	
Ahcowet SDS	I.C.I. Organics, Inc.		×																
Alacsan	Alcolac Chemical Corp.											×							
Alcolan	Robinson Wagner Co., Inc.							×											
Aldo	Glyco Chemicals, Inc.							×											
Aldosperse	Glyco Chemicals, Inc.							×											
Alkapent	Wayland Chemical Corp.					×													
Aliquat	General Mills											×							
Amber	Proctor and Gamble Co.			×															
Amerchol	American Cholesterol Products, Inc.							×											

Name	Supplier	Anionic					Nonionic				Cationic			Natural					
		Sulfates	Alkyl Sulfonic Acids and Their Salts	Carboxylic Acids and Their Salts	Alkylaryl Sulfonic Acids and Salts	Miscellaneous	Polyalkoxy Ethers	Alkoxy Esters	Alkanolamides	Miscellaneous	Amines and Amines Salts	Quaternary Amines and Their Salts	Miscellaneous	Amphoteric	Cellulose Derivatives	Water Soluble Gums	Lipids	Fluorocarbons	Silanes and Silicones
Amerlate	American Cholesterol Products, Inc.							×											
Amphocerin	Henkel International GmbH, A. H. Carnes							×											
Aquarex G	E.I. duPont deNemours & Co., Inc.		×																
Araphen	Henkel International GmbH, A.H. Carnes										×								
Arizona DRS-30, 32, 33	Arizona Chemical Co.			×															
Arlacel*	Atlas Chemical Industries, Inc.							×											
Atcor HC	Atlas Chemical Industries, Inc.										×								
Atlox	Atlas Chemical Industries, Inc.							×											
Atmos V50, 300	Atlas Chemical Industries, Inc.							×											
Atmul 80-651	Atlas Chemical Industries, Inc.							×											
Atpet 80-200	Atlas Chemical Industries, Inc.							×											
Benax 2A1	The Dow Chemical Co.				×														
Brij 30-98	Atlas Chemical Industries, Inc.						×												
Busopal NA	BASF Colors and Chemicals, Inc.				×														
Calcium Sulfonate	Swift & Co.				×														
Calimulse PRS, DMS	Pilot Chemical Co.		×																
Calsoft F90, L60	Pilot Chemical Co.				×														
Carbopol	B.F. Goodrich Chemical Co.					×													
Carolate	Robeco Chemicals, Inc.								×										
CC-16	National Lead Co.					×													

* Series of similar products.

Name	Supplier	Anionic					Nonionic				Cationic			Natural					
		Sulfates	Alkyl Sulfonic Acids and Their Salts	Carboxylic Acids and Their Salts	Alkylaryl Sulfonic Acids and Salts	Miscellaneous	Polyalkoxy Ethers	Alkoxy Ethers	Alkanolamides	Miscellaneous	Amines and Amine Salts	Quaternary Amines and Their Salts	Miscellaneous	Amphoteric	Cellulose Derivatives	Water Soluble Gums	Lipids	Fluorocarbons	Silanes and Silicones
Centrol 1P, 1F, 2F, 3P, 3F, 5F	Central Soya Co.																×		
Centrolene	Central Soya Co.																×		
Centrolex	Central Soya Co.																×		
Centromix	Central Soya Co.																×		
Centrophil	Central Soya Co.																×		
Ceralan	Robinson Wagner Co., Inc.									×									
Cetrimide	I.C.I. Organics, Inc.											×							
Chemactant A-5, LM-14	American Cholesterol Products, Inc.									×									
Chemactant AR	American Cholesterol Products, Inc.			×															
Chemactant Q	American Cholesterol Products, Inc.											×							
Chemactant S	American Cholesterol Products, Inc.						×												
Chemadene 300	Richardson Co.													×					
Concental 75	I.C.I. Organics, Inc.	×																	
Conco AAS Special # 3	Continental Chemical Co.				×														
Conco Emulsifier K	Continental Chemical Co.								×										
Conco NI 21-NI190, NI300	Continental Chemical Co.						×												
Concopal, A, AS, S, SS	Continental Chemical Co.	×																	
Comperlan	Henkel International GmbH, A.H. Carnes								×										
Complexine G	Guardian Chemical Corp.		×																
Cordon 900	Finetex, Inc.	×																	
Crodafos N-3, N-10	Croda, Inc.										×								

Name	Supplier	Anionic					Nonionic				Cationic			Natural					
		Sulfates	Alkyl Sulfonic Acids and Their Salts	Carboxylic Acids and Their Salts	Alkylaryl Sulfonic Acids and Salts	Miscellaneous	Polyalkoxy Ethers	Alkoxy Esters	Alkanolamides	Miscellaneous	Amines and Amine Salts	Quaternary Amines and Their Salts	Miscellaneous	Amphoteric	Cellulose Derivatives	Water Soluble Gums	Lipids	Fluorocarbons	Silanes and Silicones
Crodafos N-3 neutral, N-10 neutral	Croda, Inc.										×								
Dehydag Wax E	Henkel International GmbH, A.H. Carnes	×																	
Dehydazol	Henkel International GmbH, A.H. Carnes														×				
Dehydol	Henkel International GmbH, A.H. Carnes						×												
Dehydrophen	Henkel International GmbH, A.H. Carnes						×												
Dehymin	Henkel International GmbH, A.H. Carnes										×								
Dehymuls	Henkel International GmbH, A.H. Carnes							×											
Dehyquart	Henkel International GmbH, A.H. Carnes											×							
Deriphat 151, 151C, 170C, 154, 160,	General Mills													×					
Detergent 234	Dixo Company, Inc.									×									
Diam	General Mills										×								
Dispersant NI	Chevron Chemical Co., Oronite Division						×												
Dow Corning 113-202	Dow Corning Corporation																		×
Dowfax 9N 4-40	The Dow Chemical Co.						×												
Duponol	E.I. du Pont de Nemours & Co. Inc.	×																	
Ecconol B	Essential Chemicals Corp.								×										
Ecconol 70	Essential Chemicals Corp.							×											
Emulphor A Oil Soluble	BASF Colors and Chemicals, Inc.							×											

Name	Supplier	Anionic					Nonionic				Cationic			Natural					
		Sulfates	Alkyl Sulfonic Acids and Their Salts	Carboxylic Acids and Their Salts	Alkylaryl Sulfonic Acids and Salts	Miscellaneous	Polyalkoxy Ethers	Alkoxy Esters	Alkanolamides	Miscellaneous	Amines and Amine Salts	Quaternary Amines and Their Salts	Miscellaneous	Amphoteric	Cellulose Derivatives	Water Soluble Gums	Lipids	Fluorocarbons	Silanes and Silicones
Emulphor AF, P, O, O Type 1178C	BASF Colors and and Chemicals, Inc.						×												
Emulphor EL, FJ	BASF Colors and Chemicals, Inc.						×												
Emulphor FM Oil Soluble	BASF Colors and Chemicals, Inc.							×											
Emulsifier DL	Finetex, Inc.					×													
Emulsifier M85, K1	Varney Chemical Corp.												×						
Emulsifier WA	Verona-Pharma Chemical Corp.							×											
Estrex 1G 5-58	Swift & Co.			×															
Estrex P4	Swift & Co.							×											
Estrex SE 550, 750, 950	Swift & Co.							×											
Ethosperse	Glyco Chemicals, Inc.						×												
Emulgin	Henkel International GmbH, A.H. Carnes Co.						×												
EZ MUL	National Lead Co.								×										
FC 95, 98, 128, 134, FX 161	3M Company																	×	
G263, G271	Atlas Chemical Industries, Inc.	×																	
G-1045, G-3284, G-1288, G-1292, G-1295	Atlas Chemical Industries, Inc.							×											
G-1441, G-1471	Atlas Chemical Industries, Inc.						×												
G1702, G1726, G2079, G2127 G2162	Atlas Chemical Industries, Inc.							×											
G1790, G1795	Atlas Chemical Industries, Inc.						×												
G3300	Atlas Chemical Industries, Inc.				×														
G3570, G3780A	Atlas Chemical Industries, Inc.										×								
G3634A	Atlas Chemical Industries, Inc.											×							
Glycomul	Glyco Chemicals, Inc.							×											

Name	Supplier	Anionic					Nonionic				Cationic			Natural					
		Sulfates	Alkyl Sulfonic Acids and Their Salts	Carboxylic Acids and Their Salts	Alkylaryl Sulfonic Acids and Salts	Miscellaneous	Polyalkoxy Ethers	Alkoxy Esters	Alkanolamides	Miscellaneous	Amines and Amine Salts	Quaternary Amines and Their Salts	Miscellaneous	Amphoteric	Cellulose Derivatives	Water Soluble Gums	Lipids	Fluorocarbons	Silanes and Silicones
Glycosperse	Glyco Chemicals, Inc.							×											
Hartolan	Croda, Inc.									×									
Hodag 20-152	Hodag Chemical Corp.							×											
Hodag C23, 4, 5, B20	Hodag Chemical Corp.										×								
Hodag C100, 121, 130, C120, 122, 127	Hodag Chemical Corp.												×						
Hodag C128, 132	Hodag Chemical Corp.												×						
Hodag CQ 1, 2, 3, 14	Hodag Chemical Corp.											×							
Hodag DGL, DGO, DGS, EGS, EDGS, MGDT, PGS, PGL, GML, GMO, GMS, GMP, GMT, GTO, GMY, GMF, PGMS, S-35	Hodag Chemical Corp.								×										
Hodag PSML, PSMO, PSTO, PSMP, PSMS	Hodag Chemical Corp.							×											
Hodag SA	Hodag Chemical Corp.								×										
Hodag SML, SMO, STO, SMP, SMS, STS	Hodag Chemical Corp.							×											
Hodag SMO, SVO, SUS, PGMP, PGS, GMS	Hodag Chemical Corp.							×											
Hyamine	Rohm & Haas Co.											×							
ICE # 1	Glidden Co.							×											
Intexsan ABM-50, LB-MB, SB85	Washine Chemical Corp.											×							
Invermul	National Lead Co.					×													

Name	Supplier	Anionic					Nonionic				Cationic			Natural					
		Sulfates	Alkyl Sulfonic Acids and Their Salts	Carboxylic Acids and Their Salts	Alkylaryl Sulfonic Acids and Salts	Miscellaneous	Polyalkoxy Ethers	Alkoxy Esters	Alkanolamides	Miscellaneous	Amines and Amine Salts	Quaternary Amines and Their Salts	Miscellaneous	Amphoteric	Cellulose Derivatives	Water Soluble Gums	Lipids	Fluorocarbons	Silanes and Silicones
Iso-lan	Goldschmidt Chemical Corp.									×									
Isomul Extra	Isochem Corp.								×										
Isoplast M, D	Isochem Corp.							×											
Ivory	Procter and Gamble, Co.			×															
Kamar O	Finetex, Inc.	×																	
Kelecin	Textron, Inc.															×			
Kyro AC	Procter and Gamble, Co.			×															
Kyro EOB	Procter and Gamble, Co.						×												
Lanogel	Robinson Wagner Co., Inc.						×												
Lauryl Pyridinium Chloride	Hooker Chemical Corp.												×						
Lissapol 500, ACA	I.C.I. Organics, Inc.	×																	
Lissolamine	I.C.I. Organics, Inc.											×							
Lubrol	I.C.I. Organics, Inc.						×												
Maravil	Henkel International GmbH, A.H. Carnes Co.				×														
Methocel HG	The Dow Chemical Co.														×				
Methocel MC	The Dow Chemical Co.														×				
Miranol, Miranol 2MCA	Miranol Chemical Co., Inc.													×					
Miranol C2M and JEM Anhy. Acid	Miranol Chemical Co., Inc.													×					
Miranol SHD Conc.	Miranol Chemical Co., Inc.													×					
Monamides	Mona Industries, Inc.								×										
Monamine	Mona Industries, Inc.										×								
Monateric	Mona Industries, Inc.													×					

Name	Supplier	Anionic					Nonionic				Cationic			Natural					
		Sulfates	Alkyl Sulfonic Acids and Their Salts	Carboxylic Acids and Their Salts	Alkylaryl Sulfonic Acids and Salts	Miscellaneous	Polyalkoxy Ethers	Alkoxy Esters	Alkanolamides	Miscellaneous	Amines and Amine Salts	Quaternary Amines and Their Salts	Miscellaneous	Amphoteric	Cellulose Derivatives	Water Soluble Gums	Lipids	Fluorocarbons	Silanes and Silicones
Monazolines	Mona Industries, Inc.												×						
Mulsoid 815 D, M, S	Colloids, Inc.									×									
Myrj 45-53	Atlas Chemical Industries, Inc.							×											
Nacan	Allied Chemical Corp. Nat'l Aniline Div.				×														
Naccolene A Concentrate	Allied Chemical Corp. Nat'l Aniline Div.		×																
Nacconol 40 Series DBX, 355L, 60SL	Allied Chemical Corp. Nat'l, Aniline Div.				×														
Nacconol 60TL, 88SA, 98SA	Allied Chem. Corp., Nat'l. Aniline Div.				×														
Nekal AEM	BASF Colors and Chemicals, Inc.				×														
Nekanil C Highly Conc.	BASF Colors and Chemicals, Inc.						×												
Nekanil LN, 907	BASF Colors and Chemicals, Inc.						×												
Nekanil W Extra	BASF Colors and Chemicals, Inc.								×										
Nonionic E2-E30, 0-8-0-12	Hodag Chemical Corp.						×												
Nonionic L, TD, S, GR	Hodag Chemical Corp.						×												
Non-Pareil	Procter and Gamble, Co.			×															
Olate	Procter and Gammble, Co.			×															
Orvus AB	Procter and Gamble, Co.		×																
Orvus ES, K, WA	Procter and Gamble, Co.	×																	
Pegosperse	Glyco Chemicals, Inc.							×											
Petrosul	Pennsylvania Refining Co.		×																
P & G Amides	Procter and Gamble, Co.								×										
P & G Emulsifier No. 104	Procter and Gamble, Co.	×																	

Name	Supplier	Anionic					Nonionic				Cationic			Natural					
		Sulfates	Alkyl Sulfonic Acids and Their Salts	Carboxylic Acids and Their Salts	Alkylaryl Sulfonic Acids and Salts	Miscellaneous	Polyalkoxy Ethers	Alkoxy Esters	Alkanolamides	Miscellaneous	Amines and Amine Salts	Quaternary Amines and Their Salts	Miscellaneous	Amphoteric	Cellulose Dervatives	Water Soluble Gums	Lipids	Fluorocarbons	Silanes and Silicones
Plurofac	Wyandotte Chemi cals Corp.						×												
Polawax	Croda, Inc.						×												
Polychol 5-40	Croda, Inc.						×												
Promulgen	Robinson Wagner Co., Inc.						×												
Protegin X	Goldschmidt Chemical Corp.									×									
Renex 20	Atlas Chemical Industries, Inc.							×											
Renex 30-36	Atlas Chemical Industries, Inc.						×												
Ricilan	American Cholesterol Products, Inc.							×											
Rueterg 97-S	Finetex, Inc.				×														
Rueterg IPA	Finetex, Inc.	×																	
Sandrol	Clintwood Chemical Co.								×										
Saratoga	Procter and Gamble, Co.			×															
SGF	Glidden Co.							×											
Sipenol	Alcolac Chemical Corp.										×								
Sipex	Alcolac Chemical Corp.	×																	
Sipon	Alcolac Chemical Corp.	×																	
Siponate	Alcolac Chemical Corp.				×														
Siponic	Alcolac Chemical Corp.						×												
SL	Glidden Co.									×									
Sohilan	American Cholesterol Products, Inc.						×												
Solan	Croda, Inc.						×												
Soromin HS	BASF Colors and Chemicals, Inc.												×						
Span 20-85	Atlas Chemical Industries, Inc.							×											

Name	Supplier	Anionic					Nonionic				Cationic			Natural					
		Sulfates	Alkyl Sulfonic Acids and Their Salts	Carboxylic Acids and Their Salts	Alkylaryl Sulfonic Acids and Salts	Miscellaneous	Polyalkoxy Ethers	Alkoxy Esters	Alkanolamides	Miscellaneous	Amines and Amine Salts	Quaternary Amines and Their Salts	Miscellaneous	Amphoteric	Cellulose Derivatives	Water Soluble Gums	Lipids	Fluorocarbons	Silanes and Silicones
SPG 187	Glidden Co.							×											
Standamul	Standard Chemical Products						×												
Sul-Fon-ate OA5	Tennessee Corp.		×																
Sulfonated Castor Oil No. 50, 75	Swift & Co.		×																
Sulfonated Pure Neats Foot	Swift & Co.		×																
Sulfonated Soy-bean Oil	Swift & Co.		×																
Sulfonated Sperm Oil	Swift & Co.		×																
Sulfonated Red Oil	Swift & Co.		×																
Sulfopon	Henkel International GmbH, A.H. Carnes Co.	×																	
Surco 5024	Surfact Co., Inc.								×										
Surco AM-LS, TEA-LS	Surfact Co., Inc.	×																	
Surco Coconut Condensate	Surfact Co., Inc.								×										
Surfactol 318, 340, 365, 380	Baker Castor Oil Co.							×											
Surfonic LF-5, 6, 7, 16, 17	Jefferson Chemical Co.						×												
Surfonic N Series	Jefferson Chemical Co.						×												
Synotex	Dixo Company, Inc.									×									
Synthrapol	I.C.I. Organics, Inc.						×												
T-Det C-30, N	Thompson-Hayward Chem. Co.						×												
Tegacid Regular	Goldschmidt Chemical Corp.							×											
Tegacid Special	Goldschmidt Chemical Corp.			×															
Tegamine	Goldschmidt Chemical Corp.												×						

Name	Supplier	Anionic					Nonionic				Cationic			Natural					
		Sulfates	Alkyl Sulfonic Acids and Their Salts	Carboxylic Acids and Their Salts	Alkylaryl Sulfonic Acids and Salts	Miscellaneous	Polyalkoxy Ethers	Alkoxy Esters	Alkanolamides	Miscellaneous	Amines and Amine Salts	Quaternary Amines and Their Salts	Miscellaneous	Amphoteric	Cellulose Derivatives	Water Soluble Gums	Lipids	Fluorocarbons	Silanes and Silicones
Tegin P	Goldschmidt Chemical Corp.			×															
Tergitol 3A6	Union Carbide Chemicals Co.						×												
Tergitol 08, 4, 7	Union Carbide Chemicals Co.	×																	
Tergitol 12 M 10	Union Carbide Chemicals Co.									×									
Tergitol 12P Series, 15-S Series	Union Carbide Chemicals Co.						×												
Tergitol 15-S-3A, 15-S-3S	Union Carbide Chemicals Co.	×																	
Tergitol NP, TP Series, OP	Union Carbide Chemicals Co.						×												
Tergitol P-28	Union Carbide Chemicals Co.					×													
Tergitol TMN Series, X Series	Union Carbide Chemicals Co.						×												
Tetronic	Wyandotte Chemicals Corp.									×									
Texamid	Henkel International GmbH, A.H. Carnes															×			
Texapon	Henkel International GmbH, A.H. Carnes	×																	
T-Mulz	Thompson-Hayward Chemical Co.									×									
Triple C	Pro-Chem, Inc.			×															
Triton B1956	Rohm and Haas Co.									×									
Triton CF 10, 21, N-57 to 128, X-15 to 305*	Rohm and Haas Co.						×												
Triton GR	Rohm and Haas Co.		×																
Triton X151, 171, X152-172*	Rohm and Haas Co.					×													
Triton X200, X202	Rohm and Haas Co.				×														
Triton X301, 770, W-30	Rohm and Haas Co.	×																	

* Series of similar products.

Name	Supplier	Anionic					Nonionic				Cationic			Natural					
		Sulfates	Alkyl Sulfonic Acids and Their Salts	Carboxylic Acids and Their Salts	Alkylaryl Sulfonic Acids and Salts	Miscellaneous	Polyalkoxy Ethers	Alkoxy Esters	Alkanolamides	Miscellaneous	Amines and Amine Salts	Quaternary Amines and Their Salts	Miscellaneous	Amphoteric	Cellulose Derivatives	Water Soluble Gums	Lipids	Fluorocarbons	Silanes and Silicones
Triton X400	Rohm and Haas Co.											×							
Triton X700, X800	Rohm and Haas Co.					×													
Tween 20-85, 100, 280 VS	Atlas Chemical Industries, Inc.						×												
Vantoc N	I.C.I. Organics, Inc.											×							
Varamide Al	Varney Chemical Corp.							×											
Variquat K300	Varney Chemical Corp.											×							
Varonic NK	Varney Chemical Corp.						×												
Viscolan	American Cholesterol Products, Inc.							×											
Viscontran	Henkel International GmbH, A.H. Canes Co.														×				
Volpo 3, 10, 20	Croda, Inc.						×												
Waxolan	American Cholesterol Products, Inc.							×											
X-L Charge Detergent Honey	Dixo Company, Inc.					×													
Zyrol	Procter and Gamble, Co.			×															

Suppliers of Emulsifying Agents

1. Aceto Chemical Co.
2. Air Reduction Co., Inc.
3. Alabama Binder and Chemical Corp.
4. Alco Chemical Co.
5. Alcolac Chemical Corp.
6. Amalgamated Chemical Corp.
7. American Can Co., Marathon Products Division
8. American Cholesterol Products, Inc.
9. American Cyanamid Co.
10. American Dyewood, Inc.
11. American Lecithin Co.
12. Amoco Chemicals Corp.
13. Apex Chemical Co., Inc.
14. Archer Daniels Midland Co.
15. Ardmore Chemical
16. Arizona Chemical Co.
17. Arlen Chemical
18. Armour Industrial Chemical Co.
19. Arol Chemical Products
20. Atlas Chemical Industries, Inc.
21. Atlas Refinery, Inc.
22. Baker Castor Oil Co.
23. Barrett Chemical
24. BASF Colors and Chemicals, Inc.
25. Beacon Chemical Industries, Inc.
26. Berkshire Color and Chemical Co.
27. Boler Petroleum Co.
28. Burkart - Schier Chemical Co.
29. Bryant Chemical
30. Bryton Chemical Co.
31. Calgon Corp.
32. California Chemical Co.
33. Canada Packers Ltd.
34. Cargill, Inc.
35. Chemirad Corp.
36. Carlisle Chemical Works, Inc.
37. Central Soya Co.
38. Chapman Chemical Co.
39. Chemactants, Inc.
40. Ciba Chemical and Dye Co., Inc.
41. Cindet Chemicals, Inc.
42. Clintwood Chemical Co.
43. Clough Chemical Co., Ltd.
44. Colgate-Palmolive Co.
45. Colloidal Products Corp.
46. Colloids, Inc.
47. Colonial Sugars Co.
48. Commercial Solvents Corp.
49. Continental Chemical Co.
50. Crest Chemical Corp.

51. Croda, Inc.
52. Crown Chemical Corp.
53. Crown Zellerbach Corp.
54. Culver Chemical Co.
55. DePaul Chemical Co.
56. Dexter Chemical Corp.
57. Diamond Alkali Co.
58. Distillation Products Industries
59. Dixo Company, Inc.
60. C. B. Dolge Co.
61. Dominion Products, Inc.
62. The Dow Chemical Co.
63. Drew Chemical Corp.
64. DuBois Chemicals, Inc.
65. E. I. du Pont de Nemours and Co.
66. Eastern Color and Chemical Co.
67. Emery Industries, Inc.
68. Emkay Chemical Co.
69. Essential Chemicals Corp.
70. W. F. Fancourt Co.
71. Far-Best Corp.
72. Fine Laboratories, Inc
73. Fine Organics, Inc.
74. Finetex, Inc.
75. R. E. Flatow and Co., Inc.
76. Foremost Chemical Products Co.
77. Geigy Industrial Chemicals
78. General Aniline and Film Corp.
79. General Mills
80. Georgia-Pacific Corp.
81. Goldschmidt Chemical Corp.
82. Glidden Co.
83. Glyco Chemicals, Inc.
84. B. F. Goodrich Chemical Co.
85. Greenwood Textile Supply Co.
86. Guardian Chemical Corp.
87. C. P. Hall Co.
88. Hart Products Corp.
89. Henkel International Gmbh, A. H. Carnes Co., agent
90. Hercules Powder Co.
91. Hexagon Laboratories Inc.
92. Hodag Chemical Corp.
93. Hooker Chemical Corp.
94. Hope Chemical
95. E. F. Houghton and Co.
96. Humble Oil and Refining Co.
97. I. C. I. Organics, Inc.
98. International Selling
99. Intex Chemical Corp.
100. Ionac Chemical Co.
101. Isochem Corp.
102. Jefferson Chemical Co.
103. Andrew Jergens
104. Jersey State Chemicals
105. W. H. and F. Jordan, Jr., Mfg. Co.
106. Kali Mfg. Co.
107. Kalide Corp.
108. Kehew-Gradley and Co.
109. Kessler Chemical Co., Inc.
110. Knapp Products, Inc.
111. H. Kohnstamm and Co.
112. Laurel Soap Mfg.
113. Leatex Chemical Co.
114. Lever Brothers Co.
115. Leyda Oil and Chemical Co.
116. Maher Color and Chemical Co.
117. Malmstrom Color and

Chemical Corp.
118. Marden-Wild Corp.
119. Mathe Chemical Co.
120. Merix Chemical Co.
121. Metro-Atlantic, Inc.
122. M. Michel and Co.
123. Harry Miller Corp.
124. 3M Company
125. Miranol Chemical Co., Inc.
126. Mona Industries, Inc.
127. Monsanto Chemical Co.
128. Moretex Chemical Products
129. Murphy-Phoenix Oil Co.
130. National Lead Co.
131. Nopco Chemical Co.
132. Northwestern Chemical Co.
133. Nostrip Chemical Works, Inc.
134. Onyx Chemical Corp.
135. Ottol Oil Co.
136. Patent Chemicals
137. Pecks Products, Co.
138. Pennsalt Chemicals Corp.
139. Pennsylvania Refining Co.
140. Perry Brothers, Inc.
141. Pilot Chemical Co.
142. Charles Pfizer and Co.
143. Pro-Chem, Inc.
144. Proctor and Gamble, Co.
145. Proctor Chemical Co.
146. Proven Products
147. Rohm and Haas Co.
148. Relly-Whiteman-Walton Co.
149. Retzloff Chemical Co.
150. Richardson Co.
151. Robeco Chemicals, Inc.
152. Robinson Wagner Co., Inc.
153. Rozilda Laboratories
154. Ryco, Inc.
155. Sher Brothers
156. Scholler Brothers, Inc.
157. Seaboard Chemicals, Inc.
158. Shawinigan Resins Corp.
159. Shell Oil Co.
160. George F. Siddall, Co.
161. Werner G. Smith, Inc.
162. Sole Chemical Corp.
163. Solvol Chemical Co., Inc.
164. Sonneborn Chemical and Refining
165. Southern Sizing Co.
166. Fredrick A. Stresen-Reuter, Inc.
167. A. E. Staley Manufacturing Co.
168. Standard Chemical Co.
169. Standard Chemical Products
170. Stauffer Chemical Co.
171. Stepan Chemical Co.
172. Sun Chemical Corp.
173. Suffact Co., Inc.
174. Swift and Co.
175. Synthetic Chemicals, Inc.
176. Synthron, Inc.
177. Tanatex Chemical Corp.
178. Textilana Corp.
179. Textron. Inc.
180. Thompson Chemical Corp.
181. Thompson-Hayward Chemical Co.
182. Titan Chemical Products, Inc.
183. Arthur C. Trask Co.
184. Treplow Chemicals, Inc.
185. Trylon Chemicals, Inc.
186. Union Carbide Corporation (Chemicals and Silicones Divisions)
187. United Merchants and

Manufactures, Inc.
188. Universal Chemicals Corp.
189. Van Dyk and Co.
190. Varney Chemical Corp.
191. Verona-Pharma Chemical Corp.
192. Wasco Laboratories
193. Washine Chemical Corp.
194. Wayland Chemical Corp.
195. Wilson and Co., Inc.
196. Witco Chemical Co., Inc.
197. W. A. Wood, Co.
198. Woonsocket Color and Chemical Co.
199. Wyandotte Chemicals Corp.
200. General Electric Silicones Division
201. Dow Corning Corp.
202. Allied Chemical Corp, National Aniline Division
203. Chevron Chemical Co., Oronite Division
204. Tennessee Corp.

GLOSSARY

ABS—Alkyl benzene sulfonate.

Absolute (*Dynamic*) *Viscosity*—Ratio of stress to rate of shear strain.

Acid Number—Number of mg of KOH required to neutralize the free fatty acids in 1 g of fat, wax, or resin.

Adhesion—Condition in which unlike substances stick to each other.

Adhesional Wetting—Formation of a solid/liquid interface at the expense of the solid/air and the liquid/air interface.

Adhesive—Material that can hold other materials together by physical or chemical attachment at the surfaces.

Adsorbate—Material that adheres to the surface of another.

Adsorbent—Material on whose surface adsorption takes place.

Adsorption—Adhesion of a substance to the surface of a solid or liqud.

Adsorption Curve—Amount adsorbed from a solution plotted as a function of the concentration of the solute.

Aerosol—Colloidal suspension in which gas is the dispersant. Dispersion or suspension of extremely fine particles of liquid or solid in a gaseous medium.

Agglomerate—Cluster of particles adhering to one another.

Aging—Process of growing old shown by continued influence of light, temperature, external atmosphere, time, etc.

Alginates—Vegetable products made from kelp and seaweed.

Aliphatic Hydrocarbons—Straight-chain organic compounds such as the paraffins and their derivatives; e.g. propane, ethylene, acetylene.

Alkyd—Group of thermoplastic synthetic resins made from dibasic organic acids, anhydrides, alcohols, and drying oils.

Amphipathy—Simultaneous attraction and repulsion, specifically in a single molecule or ion, of one or more groups having an affinity for the phase in which the molecule or ion is dissolved together with groups that tend to be expelled by the medium.

Amphiphilic Compounds—Substances that exhibit amphipathy.

Ampholytic—Capable of acting either as a base or as an acid, depending on the nature of the surrounding medium.

Amphoteric—Reacting chemically as an acid to strong bases and as a base to strong acids.

Angstrom—Unit of wavelength equal to 10^{-10} m.

Anhydrous—Not containing water of hydration, or completely dry.

Anion—Negatively charged ion such as hydroxide (OH^-), carbonate ($CO_3^=$), phosphate ($PO_4^{\equiv}$).

Anionic Surfactants—Ionic surface-active agents in which the portion that associates with the internal phase is the anion; they include carboxylic acids, sulfuric acid esters, and sulfonic acids.

Antifoamer—Liquid of low intrinsic surface tension that prevents formation of a foam.

API—American Petroleum Institute.

Apparent Viscosity—Viscosity of a complex (non-Newtonian) fluid under given conditions.

Aromatic Hydrocarbons—Organic compounds that contain a benzenoid structure or ring. The simplest such compound is benzene, C_6H_6.

ASA—American Standards Association.

Asphalt Base Oil—Type of petroleum that yields a pitch-like residue on distillation.

Association—Combination of molecules of the same substance, resulting in multiple molecules.

ASTM—American Society for Testing and Materials.

ASTM Time Tube—Viscometer of the rising-bubble type.

Atlantic Viscometer—Apparatus used to determine the kinematic viscosities of transparent Newtonian liquids of 0.6-5000 cS.

Attrition—Pulverization through abrasion.

Baffle—Partition or plate that changes the direction or restricts the cross section of a fluid, thus increasing velocity or turbulence.

Base Stock—Blend of materials that makes up the bulk of a product.

Batch—Fixed unit charge in the processing of materials.

Bentonite—A colloidal clay used as an adsorbent.

Benzin (*Petroleum Benzin*)—A refined light naphtha meeting USP specifications.

Binder—Nonvolatile portion of coating vehicle in the film-forming ingredient that is used to bind the pigment particles together.

Biodegradability—Susceptiblity of a chemical compound to depolymerization by the action of biological agents.

Bingham Plastic—A substance that will not flow until its yield value is reached; thereafter, it flows like a normal liquid.

Binodal Curve—Curve in a triangular isothermal equilibrium plot that connects miscibility points.

Biocolloid—Lyophilic colloidal system found in living organisms.

Bitumen—Mixture of natural and pyrogenous hydrocarbons that is soluble in carbon disulfide.

Bleeding—Separation of the oil constituent of a grease from the base.

Blending—Mixing of several components to obtain a homogeneous material.

Bloom—Appearance of an emulsion as it is poured into a large amount of its continuous phase.

Blowing—Injection of compressed air into tanks to agitate and mix the contents; the act of foaming a system.

Blowing Agent—Material used to provide the gas needed for foaming operations.

Bond—Linkage between atoms in a molecule.

Branched—Condition in the molecular structure of polymers where the polymer has side chains attached to the main chain.

Breakdown Time (gasoline)—Time elapsed before gasoline forms gums in storage.

Brookfield Viscometer—Instrument that measures the viscosity of a liquid as a function of the torque produced by rotating a spindle at fixed rpm in a liquid.

Brownian Movement—Random movement of colloidal-size particles that arises from collisions between the particles and molecules of the fluid medium in which they are suspended.

B Stage of Resin—Partial cure of a resinous polymer in which it is more viscous with higher molecular weight, being insoluble but plastic and fusible.

Buffer Action—Property of certain solutions of resisting changes in pH upon the addition of acid or alkali.

Buffer—Substance that causes a resistance to change in pH when added to a solution.

Bulk Density—Weight per unit volume of a quantity of solid particles; depends on packing density.

Bunker Fuel—Fuel for marine boilers.

Calendering—Subjecting a material to pressure between two or more counter-rotating rollers.

Cannon-Fenske Viscometer—Modified Ostwald-type viscometer used to determine the kinematic viscosity of transparent Newtonian liquids (0.4-16,000 cS).

Capillary Action—Phenomenon observed on the surface of liquids due to unbalanced molecular attraction at the liquid boundary.

Capillary Tube Viscometer—Instrument for the measurement of viscosity as a function of the time required for a certain amount of fluid to flow through a small-bore tube of known diameter and length under a known constant pressure difference. Assuming laminar flow, the viscosity is calculated from the Hagen-Poiseuille law.

Capillary Value of Adsorbent—Difference between saturation and retentivity value. Amount of solvent held loosely in pores.

Catalyst—Substance that alters the rate of chemical change and remains unchanged at the end of a reaction.

Catalyst Activity—Measure of the relative performance of a catalyst in promoting the reaction of a material to give a desired performance.

Cataphoresis—Particle migration in a colloidal solution between electrodes exhibiting a potential difference.

Cation—Positively charged ion such as Na^+, K^+, NH_4^+.

Cationic Surfactants—Ionic surface-active agents in which the portion that associates with the internal phase is the cation. They include simple amine salts, quaternary ammonium salts, amino imides and imidazolines. Cationic surfactants often have germicidal, anticorrosive, and antistatic properties.

Centipoise (cP)—Unit of viscosity equal to 1/100 dyne/sec^2/cm^2.

Channeling—Physical parting or separating of the body of a lubricant.

Chemical Change—Change in which new substances with new properties are obtained from a chemical composition or decomposition of a substance.

Chemisorption—Adsorption process in which a polar group is bound to certain groups in a surface by chemical linkages.

Chiller—Cooler for separating wax from petroleum fractions.

Clay Treatment—Adsorption process used in oil refining.

Cloud Point—Temperature at which solutions become cloudy on cooling or heating.

Coagulation—Process in which the small particles in an emulsion are combined in agglomerates but still retain their character as individual particles.

Coalescence—Process by which small particles (droplets) in an emulsion combine to form larger particles (drops).

Coarse Emulsion—Emulsion that contains large globules.

Cohesion—Force of attraction of like molecules for one another within a substance.

Cold Pressing—Separation of a solid petroleum product from a liquid petroleum medium by means of a filter press.

Collector—Chemical additive controlling the wettability of ores in a flotation process.

Colloid—Finely divided substance 0.001-1 μ in diameter. The physical properties of a colloid are controlled by the surface forces.

Colloidal Metals—Colloidal suspension of metals.

Colloid Mill—Machine used for the prepation of emulsions of fine particle size. The emulsion is forced through the adjustable small annular space between a stator and a rapidly revolving rotor.

Colloidal Suspension—Two-phase system having small dispersed particles suspended in a dispersant.

Complex Liquid—Liquid in which the rate of shear is not proportional to the shearing stress; a non-Newtonian liquid.

Complex Mixture—Mixture of more than three components.

Compound—Two or more elements chemically united in definite percentages by weight that are separable only by chemical means.

Compound Oils—Oils made by mixing minerals oils with other oils, vegetable or animal.

Concentration of Emulsion—Volume percent (or volume fraction) of the internal or dispersed phase.

Concentration of Emulsifying Agent—Amount of surface-active agent in the emulsion. May be based on the volume or weight of the total emulsion or on one phase.

Conjugate Layers—Coexisting layers in an extraction—the solvent and the diluent.

Conradson Test—Method for the determination of the percentage by weight of carbonization residue left from burning a petroleum product according to an ASTM standard method.

Consistency—Reistance of a fluid to deformation. For simple (Newtonian) fluids the consistency is identical with the viscosity, for complex (non-Newtonian) fluids, identical with apparent viscosity.

Contact Surface—Area of a phase interface.

Continental Method—Procedure for preparing emulsions whereby the oil and surfactant(s) are throughly mixed and water is added to the mixture. As progressively more water is added, the emulsion will invert and water will become the continuous phase.

Continuous Phase—External phase of an emulsion.

Convection—Natural or forced motion in a fluid induced by heat or the action of gravity.

Copolymer—Compound resulting from the chemical reaction of two different monomers or polymers with each other.

Couette Viscometer—Rotational type of instrument for the measurement of viscosity. A fluid is placed in the annular space between two concentric cylinders, one of which is rotated. The measurement of torque and velocity gradient can be correlated with viscosity.

Coulomb Viscometer—Instrument that measures the damping effect produced by a fluid upon a vibrating reed.

Covalent Molecule—Molecule in which the bond between two atoms is a shared electron pair.

Creaming—Separation of an emulsion into two distinct phases, one of which has a higher concentration of the dispersed phase than the original emulsion; the other contains more of the continuous phase.

Critical Micelle Concentration (CMC)—Saturation point of a surfactant in a water system. The CMC is observable by discontinuities in the curves of physical properties of the system as a function of the amount of surfactant added.

Critical Quantity—Minimum or maximum amount of a blended or homogenized material needed to obtain a desired result.

Critical Surface Tension—Surface tension of a solid equal to that of the liquid of highest surface tension that will wet the solid.

Critical Velocity—Transitional velocity at which the flow of a fluid changes from laminar to turbulent.

Crosslinks—Bonds linking molecular chains.

Cut-Back Product—Material that has been diluted by adding a solvent.

Dead Oils—Tar distillates that are heavier than water.

Deblooming—Bleaching of oils to reduce fluorescence.

Deflocculate—Converse of clumping or clustering of particles.

Defoamer—Substance of low intrinsic surface tension that reduces or destroys frothing.

Denaturing—Making a product unfit for human consumption, as by the addition of chemicals influencing the taste or toxicity.

Densimeter—Instrument for measuring the density or specific gravity of liquids.

Detergent—Surface-active material or combination of surfactants designed for removal of unwanted contamination from the surface of an article.

Dialysis—Separation of colloids from dissolved substances by selective diffusion through a semipermeable membrane.

Diffusion—Mixing of molecules or atoms by random molecular or atomic motion.

Dilatant—Increasing instantaneously the apparent viscosity of a fluid with increasing rate of shear.

Diluent—Diluting agent.

Dilution—Act of reducing the proportion of a substance in a solution by addition of more solvent.

Dilution Ratio—Volume of solute divided by volume of solvent.

Dipole—Molecule that has an electrical moment (charge separated by a distance).

Dispersed Phase—Dissolved or suspended phase in a colloidal solution or suspension.

Dispersing Agent—Surfactant that is used to separate, or keep separated, aggregations of particles by reducing the cohesive attraction between the particles.

Dispersion—Physical, usually temporary, mixture of two insoluble phases.

Dispersing Medium—Continuous phase of a dispersion.

Dissociation—Breaking up a molecule or associated species into smaller entities.

Dissolving—Formation of a solution by dispersion of one material (solute) at a molecular (or less) level in another material (solvent).

Distribution Ratio—Ratio of concentrations of a dissolved subtance in two immiscible solvents.

Double Layer Potential—Electrical potential difference between a substance and a solvent.

Dropping Point—Temperature at which a grease passes from a semi-solid to a fluid state.

Dual Emulsion—Emulsion in which both the internal and external phase contain portions of the opposite phase instead of being homogeneous.

Dynamic Similarity—Condition in which all corresponding forces between two systems are proportional.

Dynamics—Behavior of bodies under the action of forces that produce changes of motion in them.

Dynamic Viscosity—An index of molecular friction within a fluid that is proportional to the rate of deformation and the viscosity coefficient of a fluid (absolute viscosity).

Ebullioscopy—Determination of the concentration of materials in a solution based upon boiling point deviations.

Eddy Viscosity—Turbulent viscosity based upon ratio of shear stress to a velocity gradient.

Elastic Fluid—Fluid in which elastic stresses and hydrostatic pressures are large compared with viscous stresses.

Elasticity—Recovery of the original size and shape of a material after deformation.

Electrophoresis—Migration of the electrically charged particles toward the oppositely charged electrode.

Electroviscous effect—Lyophobic particles that bear an electrical charge and therefore exhibit a viscosity exceeding that of a similar system of uncharged particles.

Emulsifying—Process of uniformly dispersing one mutually insoluble material in another.

Emulsifying Agent—Surface-active agent that is primarily used to effect the stability of an emulsion system. Emulsifying agents may be ionic (anionic, cationic, amphoteric) or nonionic in solution.

Emulsion—Beta-glucose.

Emulsion—System consisting of two incompletely miscible liquids, one being dispersed as finite globules in the other. A small amount of a third substance may render the dispersion stable. The liquid broken up into globules is the dispersed (discontinuous) phase; the surrounding liquid is the external (continuous) phase.

Emulsion Polymerization—Polymerization of a water insoluble monomer that takes place in the presence of water to form a latex. The monomer is usually emulsified and polymerization occurs within the emusion system.

Emulsoid—Colloidal system in which there is a strong attraction between the disperse substance and the dispersing liquid.

Emulsoid Sol—Colloidal suspension in which both dispersion medium and dispersed phase are both liquids.

Endothermic Process—Process that absorbs heat.

Engler Degree—Measurement of kinematic viscosity that is obtained by dividing the outflow time of a specified amount of liquid through a standard orifice by the outflow time of water at 20°C.

English Method—Procedure for the preparing of an emulsion whereby a small portion of the water and the surfactants are mixed. The oil and remaining water are then added in successive small portions.

Eucolloid—Colloid composed of relatively large particles (larger than 0.25 μ).

Exothermic Process—Process that liberates heat.

External Phase—Dispersing medium of a colloidal suspension; the continuous phase of an emulsion.

Extrusion—The process of forcing a material in plastic condition through an orifice.

Falling-Body Viscometer—A device that utilizes a sphere, falling through a mass of the test fluid inside a cylindrical tube. The time for the sphere to fall a certain distance is measured and correlated with viscosity.

False Body—Reversible work-softening that causes a high apparent viscosity of a material.

False Gel—Unstabilized gel that breaks down upon standing.

Fire Point—Lowest temperature of an oil at which it will ignite and continue to burn for at least 5 seconds.

Flash Point (*Cleveland Open-Cup Test*)—Lowest temperature of an oil at which it gives off vapors that will ignite when a small flame is passed over the surface of the oil.

Fitzsimmons Viscometer—Device that determines the kinematic viscosities of transparent Newtonian liquids in the 0.6-1200 cS range.

Flocculation—Process in which the disperse particles in a liquid medium display a tendency for mutual adhesion.

Flotation—Ore-dressing method in which surface-active additives cause one of the constituents to float to the surface in the froth produced by aeration and agitation.

Fluid—Substance that cannot sustain shear forces when in static equilibrium. Real fluids in motion sustain shear forces because of the property of viscosity.

Fluidity—Reciprocal of viscosity, expressed in inverse poise.

Fluxing—Thinning of heavy oil by the addition of a more fluid oil.

Foaming Agent—Surface-active material that is used specifically to form a dispersion of a gas in a liquid or solid medium.

Force-Type Viscometer—Instrument that measures force at constant shear rate. Units of force can be converted directly into viscosity units.

Formula Weight—Sum of the atomic weights of all the atoms in an empirical formula.

Friction Factor—Used in the Fanning equation to compute the friction loss through a length of tube and is a function of Reynolds number.

Frothing Agent—Surface-active material that is used specifically to stabilize the dispersion of a gas in a liquid or solid medium.

Galling—Metallic binding or seizure under load and relative motion.

Gardner Mobilometer—Instrument used to determine the thixotropy of a material.

Gel—Semisolid system that consists of a solid held in a liquid; a more solid form than a sol.

Gelometer—Instrument used to measure the time required for a fluid to gel.

Gibbs Absorption Law—When the concentration of a substance in the interfacial layers between two phases exceeds that in the bulk of either phase, the substance is absorbed at the interface.

Gouy Diffuse Double Layer—The gradual, exponential electrical potential drop from the interface of an emulsion into the bulk of the phase.

Gum—Class of colloidal substances that is exuded by plants.

Hallikainen Shell—Type of viscometer that yields continuous values of a pressure differential maintained in a capillary tube for the determination of fluid viscosity.

Hegman Gage—Arbitrary scale for comparing fineness of grind, expressed in "mesh."

Helmholtz Double Layer—The sharp and continuous decrease into the bulk of the fluid of the electrical potential of a single layer of oriented, ionized surfactant molecules at the interface. This effect can exist on both sides of the interface in an emulsion.

HLB Number (*Hydrophile-Lipophile Balance*)—System for the selection of emulsifying agents. *Lipophilic* (nonpolar) emulsifiers have HLB numbers below 9; *hydrophilic* (polar) have HLB numbers above 11. The HLB of an emulsifier (or blend of emulsifiers) may be related to its performance in an emulsion system.

Holdup—Amount of liquid retained by a piece of equipment when the equipment is emptied.

Homogeneous Fluid—Substance that is identical throughout.

Homogenization—The process of making incompatible or immiscible components into a stabilized uniform suspension in a liquid medium.

Homogenizer—Mixing machine used for the preparation of emulsions of fine particle size. The emulsion is forced at high pressure through the annular space between an adjustable valve and its seat.

Homopolymer—Polymer that is produced from a single monomer species.

Hydrometer Densimeter—Device used for the measurement of specific gravities.

Hydrate—Crystallized substance that contains water of hydration.

Hydration—Attachment of water molecules to ions or molecules.

Hydrogel—Colloidal gel in which water is the solvent.

Hydrolysis—Splitting of a chemical bond(s) by the addition of water.

Hydrophilic—Attracted to water; water-soluble.

Hydrotopy—Effect of large additions of various substances to water in increasing the solubility of other substances therein.

Ideal Fluid (*Perfect Fluid*)—Hypothetical liquid (or gas) that offers no resistance to shear and has zero consistency.

Impeller—Rotating member of an agitating or pumping device.

Implosion—Inward collapse.

Inhibiter—Substance that prevents or slows down a chemical reaction.

Initiator—Substance that starts a polymerization reaction.

Interfacial Film—A film of material (usually liquid) that forms at the point of contact of two immiscible liquids. The film may prevent coalescence.

Interfacial Viscometer—Instrument used for the measurement of the physical properties of surface and interfacial films.

Internal Phase—Dispersed or discontinuous phase of an emulsion.

Intumesce—Foaming, swelling, or bubbling that may be a result of heating.

Inversion—Transposition of the two phases of an emulsion (internal phase becomes external phase and vice-versa).

Ion—Electrically charged portion of matter of atomic or molecular dimensions.

Ionic Surfactants—Surfactants that dissociate in solution into an organic lipophilic group and a hydrophilic group. The molecules produce two ions, one positively charged (cation), the other negatively charged (anion).

Ionic Solutions—Solutions of substances such as acids, bases, and salts that dissociate in solution into charged particles called ions.

Iso—Prefix meaning equal.

Isoelectric Point—pH value of a hydrophilic sol (electrolyte) at which it will not migrate in an electrical field.

Isotropic Fluid—Fluid whose properties are independent of the rotation of the axis of reference along which the properties are measured.

Key Material—In a blend of two or more materials, material whose uniform dispersion is most important to the success of the operation.

Kinematic Eddy Viscosity—Ratio of eddy viscosity to density.

Kinematic Viscosity—Ratio of viscosity to density; the unit of kinematic viscosity is the stoke.

Kinematics—Motion of bodies without reference to the forces producing the motion.

Krebs Units—Arbitrary scale used for the comparison of viscosities.

Laminar Flow—Streamline flow of a fluid. The mass of fluid advances in separate sheets with simple shear existing at the surface of contact of different layers if the layers move at different speeds.

Lantz-Zeitfuchs Viscometer—Reverse-flow viscometer that permits the measurement of kinematic viscosities of opaque Newtonian liquids of 60-120,000 cS.

Latex—Any stable dispersion of insoluble polymer particles in a water system. Generally an emulsion of a high molecular weight polymer in water, often formed by emulsion polymerization of the monomer(s).

Leaching—Washing operation that removes a certain amount of dissolved substance(s) from a material.

Levigation—Grinding of a solid to a fine smooth powder while in moist condition.

Lipophilic—Attracted to oil; oil-soluble.

Liquid—Substance that undergoes continuous deformation when subjected to shearing stress.

Livering—Coagulation of a paint or enamel into a viscous rubbery mass.

Lubricating Greases—Solid or semifluid products of the dispersion of a thickening agent in a liquid lubricant that may contain ingredients to impart special properties.

Lyophilic—Pertaining to a colloidal system of the emulsoid type.

Lyophobic—Pertaining to a colloidal system of the suspensoid type.

Lyophobic Colloid—Irreversible colloid; the disperse phase when removed from suspension does not return to the colloidal state upon simple remixing.

Lypophilic Colloid—Reversible colloid; the dispersed phase when removed from suspension returns to the colloidal state on remixing with the dispersion medium. The suspension is a reversible sol.

MacMichael Viscometer—Viscometer that consists of an outer cylinder rotated by a variable-speed motor; the measured torque exerted on an inner cylinder is a function of the viscosity of the test fluid. It is suitable for non-Newtonian suspensions.

Macromolecule—Very large molecule, generally of polymer.

Master Batch—Blend of several minor ingredients that is used later in the compounding of a product.

Master Viscometer—Glass viscometer (capillary) with a liquid driving head of 400 mm or better, designed to minimize errors due to surface tension, kinetic energy, and capillary end effects.

Maximization—Concerted effort to increase the efficiency of a process through the most appropriate combination of instruments and techniques.

Maxwellian Fluid—Viscous fluid in which the stress-strain relationship includes the relaxation effect (time of relaxation of elastic stresses set up by sudden deformation).

Mayonnaise Technique—Emulsification procedure in which the ingredients are added alternately.

Micron—Unit of length, equals 1 micrometer or 0.001 mm.

Milling—Final "polishing" operation (prior to packaging) to improve the stability of a grease by modifying its gel structure.

Miscella—Solvent-oil solution obtained in the extraction of certain oil seed and containing mucilaginous materials that may be difficult to separate.

Micelle—Small aggregation of surfactant molecules in a suspension. If water is the dispersing medium, the molecules are orientated with the hydrophilic end toward the water phase and the lipophilic ends together.

Mixture—Material composed of two or more substances, each of which retains its own characteristic properties.

Modifier—Additive that changes the nature of a product or the process by which it is produces.

Modulus of Elasticity—Ratio of stress to strain in a material that is elastically deformed.

Molar Volume—Molecular weight of a substance divided by its density.

Molecule—The smallest theoretical quantity of a material that retains the properties exhibited by the material.

Molecular Association—Aggregation of single molecules into complexes containing two or more molecules.

Molecular weight—Sum of the atomic weights of all the atoms in a molecule.

Mole—Mass numerically equal to the sum of the atomic weights of all the atoms in a molecule.

Monomer—Simple form of a compound that is the building block of a polymer.

Multiple Emulsion—Emulsion in which the dispersed phase contains globules of the continuous phase.

Natural Gums—Plant (usually tree) exudations such as gum arabic, gum tragacanth, gum karaya, and guar that are high-polymer saccharides.

NF—National Formulary

Neutralization Number—mg of KOH required to neutralize the acid in 1 g of oil.

Newton Viscometer—Instrument that measures the resistance of a fluid to motion between two parallel sliding plates. A Newton viscometer is often used on high-viscocity materials.

Newtonian Fluids—Fluids whose viscosity is independent of the rate of shear. Pure liquids, true solutions, and dilute suspensions are generally Newtonian.

Newton's Law of Similarity—Material systems are dynamically similar when the ratio of forces acting upon corresponding mass elements at corresponding times is constant. Such forces include pressure, inertia, gravity, viscosity, elasticity, and surface tension.

Nip—Distance between rolls in mill or calender.

Nonelectrolyte—Substance whose water solution will not noticeably conduct an electric current.

Nonionic Surfactants—Covalent emulsifying agents that will not ionize in solution; they include alkyl ethers, alkylaryl ethers, thio-ethers, esters, and amides.

Non-Newtonian Suspensions—In viscous flow, the viscosity of a non-Newtonian suspension is a function of the rate or shear and duration of the shear. Non-Newtonian classes of materials include Bingham plastics, pseudoplastics, dilatants, thixotropes, and rheopectics.

Nonpolar Liquids—Liquids whose molecules possess little or no dipole moment.

Normal Hydrocarbons—Hydrocarbons with straight, not branched, molecular chains.

Oil—Common shorthand in emulsion technology referring to the water-insoluble phase of an emulsion. The oil phase may be a flourocarbon, silicone, or wax, in addition to or instead of a conventional oil.

Organosol—Colloidal suspension in which the continuous phase is an organic compound.

Osmosis—Diffusion between two miscible fluids separated by a permeable wall.

O/W Emulsion—Oil-in-water emulsion; oil is the internal phase, water the external phase. An O/W emulsion is dispersible (dilutable) in water, but not in oil.

Parasitic Reaction—Side reaction that may interfere with a process and reduce its yield.

Partial Pressure—Portion of the total pressure of a system that is the pressure contribution of a specific component of the gas mixture.

Penetrant—Surface-active agent used in the texile industry to aid dyeing and sizing operations.

Penetrometer—Instrument for determining the consistency of high-viscosity materials such as asphalts, greases, waxes.

Peptization—Dispersion of a substance into particles of colloidal size, caused by addition of a peptizing agent.

Peptizing Agent—A protective colloid.

Percolation—Extraction process, usually of a liquid acting upon a solid.

Perfect Fluid—Fluid of zero viscosity.

Permanent Emulsion—Emulsion in which the droplets of the dispersed phase do not coalesce or coagulate in time; a stable emulsion.

Phase Interface—Boundary surface between two phases or films.

Phase Rule—For a heterogeneous system in equilibrium, the sum of the number of phases plus the number of degrees of freedom equals the number of components plus two.

pH Value—Negative logarithm of the hydrogen ion concentration and hence the acidity or alkalinity of a solution. The scale ranges from 1 (highly acidic) through 7 (neutral) to 14 (highly basic.)

Phase-Volume Ratio—Ratio of the volume of the internal phase to that of the external phase in an emulsion.

Photomicrograph—Photograph of the emulsion that is used for particle size determinations or other visual observations of the emulsion. Agglomeration can be seen if the particles are stabilized by the addition of a thickener, e.g. gelatin, to the continuous phase.

Physical Change—Change in which the identifying chemical properties of a substance remain unchanged.

Pigment—Finely divided insoluble substance that imparts color to the material to which it is added.

Pipeline Mixer—Device for continuous mixing and/or homogenizing within a transfer pipe.

Pitting—Spot corrosion that is caused by chemical or galvanic effects.

Plastic Deformation—Change in the dimensions of a material load that is not recovered when the load is removed.

Plasticizer—Chemical agent added to plastics to make them softer and more flexible.

Plastisol—Suspension of resin(s) in plasticizer(s).

Plastic Solid—Substance that does not deform under a shearing stress until the stress attains the yield stress, at which point the solid deforms permanently.

Plug Flow—Flow characterized by an almost square velocity profile. Under certain pressure conditions prevailing in closed tubes, mass in the middle of the pipe moves like a solid plug, with laminar flow along the wall.

Poise—Unit of viscosity, equivalent to 1 dyne/sec/cm^2.

Poiseuille's Equation—Relation between liquid flow through a tube, pressure, and length and cross-section of path, assuming uniform streamline flow.

Poiseuille Viscometer—Instrument to measure viscosity based on outflow measurement of fluids. Commercial Poiseuille viscometers include Engler, Barbey, Redwood, and Saybolt units.

Polar Bond—Essentially covalent chemical bond that results from the unequal sharing of valence electrons between atoms.

Polar Compound—Substance that possesses a permanent electric moment.

Polar Liquids—Liquids whose molecules possess a dipole moment, such as alcohols, acids, aldehydes, etc.

Polydisperse Systems—Systems that consist of various size particles or droplets.

Polymer—Compound formed by polymerization, characterized by a high molecular weight (usually over 1000) and a repeating chemical structure.

Polymerization—Formation of large molecules from smaller ones (monomers).

Polyphase Emulsion—Mixed, multiple, or dual emulsion that has no well-defined internal and external phases; each phase contains droplets of the other phase.

Pour Point—Lowest temperature at which an oil will pour.

Precipitate—Substance separated from a solution, suspension, dispersion, or emulsion as a result of chemical or physical change.

Pressure Drop—Difference in force on unit areas on the two sides of a barrier.

Protective Agent—Colloidal substance that stabilizes the suspension of the particles when adsorbed on suspended particles.

Protective Colloid—Material that stabilizes a particular disperse system against coagulation or coalescence. It is usually a lyophilic colloid that forms a protective film (by adsorption) around each particle.

Pseudoplasticity—Property of a fluid whose consistency decreases instantly with increasing rate of shear.

Pseudostable—State of a material that causes a change of its physical properties over a period of time.

Quiescent Flow—Condition existing when the degree of turbulence is less than that indicated by the Reynolds number.

Random Distribution—Arrangement of particles in a mixture so that the probability of a single particle being at any one location is the same as that of its being at any other location.

Rao Birefringence Viscometer—Couette-type viscometer that is used for low-viscosity fluids.

Relative Viscosity—Viscosity of a liquid in terms of the viscosity of some standard liquid, such as water at 20°C.

Relative Volatility—Volatility of one component divided by volatility of the other components; the pressure ratio for components following Raoult's law.

Resolving Power—Ability of a microscope to produce separate images of particles very close together.

Reynolds Number (*Re*, *dimensionless*)—Measure of the degree of turbulence of a fluid flowing in a pipe: diameter times velocity times density and divided by absolute viscosity.

Rhe—Unit of fluidity; 1 rhe $=1$ poise^{-1}.

Rheological Properties—Flow properties of a system, e.g. plastic flow, thixotropy, and pseudoplasticity.

Rheology—Study of the deformation and flow of matter.

Rheopectic—Pertaining to a fluid whose apparent viscosity increases with time to some maximum value at any constant rate of shear.

Rigidity—Reciprocal of mobility.

River-Water Die-Away Test—Proposed test to determine the biodegradability of surface-active agents.

Roller Mill—Two or more rolls, usually revolving at different speeds, between which the materials to be mixed are passed.

Rotational Flow—Flow that exhibits appreciable vorticity.

ROV (*Refined Oil Viscosity*)—Viscosity of lamp oils and napthas as measured by a Saybolt Thermo-viscometer.

Rupture Stress—Shear stress.

Salting Out—Precipitation of a material from its dispersion.

Saponification—The process of hydrolysis of fats or oils by an alkali of a fluid to form soap.

Saybolt Universal Seconds (*SUS*)—Kinematic viscosity of a fluid as determined by the time in seconds required for 60 cm^3 of liquid to flow through a standard orifice.

Saybolt Viscometer—Transpiration type of viscometer primarily used to determine viscosity of petroleum products.

Scattering—Deflection of light rays by particles of suspended matter.

Sedimentation—Spontaneous separation of two or more materials (liquids, solids, or gases).

Segregation—Separation of the constituents of a mixture, e.g. by electrostatic charges.

Seizing—Metallic freezing together of materials under load.

Shake Flask Test—Screening test for the determination of biodegradability, proposed by the Soap and Detergent Association.

Shear Stability—Ability of a grease to withstand repeated working with minimum change in structure or consistency.

Shear Viscometer—Instrument for measuring viscosities of complex fluids at different shear rates.

Shelf Life—Period of time during which an undisturbed emulsion retains its type and essential properties.

Shock Chilling—Addition of a cold fluid to a hot fluid in order to obtain rapid cooling of the blend to a given temperature.

Sil Viscometer—Instrument used for the determination of viscosities of Newtonian fluids in the 0.6-10,000 cS range.

Simple Liquid—Liquid in which the rate of shear is proportional to the shearing stress. The constant ratio of shearing stress to rate of shear of a simple liquid is the viscosity of the liquid.

Soap—Metallic salt of a fatty acid.

Slip—Displacement loss in a pump that is affected by the viscosity of the liquid being pumped and the pressure differential.

Solid—Substance that undergoes permanent deformation only when subjected to shearing stress in excess of some finite value characteristic of the substance (yield stress).

Solids—Active ingredients of a mixture. In the emulsion the "solids" includes the internal phase and the emulsifiers, stabilizers etc; all of the system except the continuous phase or solvent.

Solubility of an Emulsion—Dispersibility of an emulsion as determined by the nature of the continuous phase.

Solute—Substance that is dissolved in a solvent; the substance in lesser concentration in a solution.

Solution—Homogeneous mixture of two or more substances. It is characterized by an ionic or molecular-level subdivision of the components that will not become heterogeneous upon standing.

Solvation—Adsorption of molecules of a dispersing liquid by the dispersed particles.

Solvent—Constituent of a solution present in the larger amount.

Space Velocity—Product of the feed rate times specific volume of feed divided by volume of vessel used in a converting operation, expressed in volume processed per unit time. It is the reciprocal of hypothetical retention time.

Specific surface—Surface area of a substance per unit volume.

Specific Viscosity—Relative viscosity of a fluid.

Specific Volume—Reciprocal of density; volume per unit mass, as cm^3 /kg, ft^3/1b, etc.

Speed Velocity—see *Space Velocity.*

Spreading Coefficient (*Harkins*)—Number describing the decrease in the total interfacial energy of an aqueous system by the addition of a specific surface active substance.

Stable Emulsion—Emulsion that retains its form and condition and in which the particles of the internal phase do not coalesce.

Stabilizer—Ingredient that is added to a system to maintain the physical and chemical properties of the product throughout its processing and service life.

Stagnation Point—Point at which the fluid velocity of a system is zero.

Standard Deviation—Statistical measure of the scattering of data from the average; equal to the root mean square of the individual deviations from average.

Stern Diffuse Double Layer—Presence of a sharp Helmholtz layer of ionic thickness at the interface followed by a Gouy exponential electrical potential drop into the bulk of the fluid. This effect is two-sided; it exists on both sides of the interface in an emulsion.

Stoke—unit of kinematic viscosity, symbol S.

Stokes' Law—A small sphere falling under the action of gravity through a viscous medium reaches a constant velocity that is a function of the radius of the sphere, its density, the density of the medium, and its viscosity coefficient.

Stormer Viscometer—Viscometer in which the fluid is placed between two cylinders. The inner cylinder rotates by the action of a falling weight. The rotational speed is determined by measuring the time required for a definite number of revolutions of the inner cylinder. The torque is determined from the weight and radius of the pulley. The viscosity of the fluid is a function of the rotational speed and torque.

STP—Standard Temperature and Pressure: 0°C, 760 mm pressure.

Strain Deformation—Change in the dimensions of a body under stress.

Stratification—Formation of layers in a fluid mixture.

Streamline—Line that gives the velocity direction of a fluid (number of particles) at each point along the line.

Streamline Flow—Flow where fluid particles move along the streamlines.

Stress—Internal force exerted by one part of a body upon an adjoining part.

Structural Color—Color due to the physical structure of the mass and not dependent upon the electronic configuration of the substance.

Substance—Any form of matter that is homogeneous throughout, i.e. solid, liquid, gas.

Supersaturated Solution—One that contains more solute in solution than is present in a saturated solution of the same substance at the same temperature and pressure. A supersaturated solution is usually formed by reducing the temperature or removing some of the solvent from a saturated solution.

Surface-Active Agent—Substance that affects the surface tension of a liquid. They include emulsifying agents, detergents, suspending agents, wetting agents, etc.

Surfactant—Surface-active agent.

Suspensoid—Colloidal suspension in which there is little attraction between the dispersed substance and the dispersing liquid; a lyophobic colloid.

Syneresis—Exuding of small amounts of liquids from gels by contraction.

Thermoplastic—Resin that can be readily softened and resoftened by heat.

Thickner—Device that produces separation of solids from liquids by sedimentation; material that increases the viscosity of a fluid.

Thick Phase—Increase in viscosity found in an emulsion before inversion. The close packing of the internal phase results in a high viscosity. The thick phase is often used to maximize the shear on an emulsion by mixing the emulsion into a thick phase or causing inversion by mixing the thick phase.

Thixotropy—Reversible gel-sol-gel transition that is characterized by a reduction in viscosity upon the application of shearing stress.

Time-Measuring Viscometer—Instrument measuring the time needed for effiux of a liquid through an opening, rise of an air bubble, fall of a steel ball, revolution of a paddle driven by a constant torque, or slide of a plate.

Torr—Unit of pressure, same as millimeter of mercury.

Turbidimeter—Instrument that measures the reduction in light transmission caused by interposing a suspension between light source and observer.

Turbulence—Flow caused by superimposing irregular currents on a uniform flow.

Turbulent Flow—Motion of a fluid in which its velocity at a fixed point fluctuates randomly with time.

Tyndall Effect—Light-scattering that is visible when a beam of coherent light passes through a colloidal system.

Ubbelohde Viscometer—Suspended level instrument that determines kinematic viscosities of transparent Newtonian liquids in the 2-

10,000 cS range.

Ultrasonic Emulsifier—Machine that is used for the preparation of fine-particle-size emulsions. The emulsion is forced at high velocity against the edge of a blade. The acoustical waves of the vibration of the blade produce cavitation, which accomplishes the emulsification.

Unstable Emulsions—Those emulsions that separate into two immiscible phases, or otherwise lose homogenity.

USP—United States Pharmacopeia.

Valence—Number of electrons gained, lost, or shared by an atom in bonding to one or more other atoms.

Velocity of Approach—Average velocity of fluid in a duct ahead of the point considered.

Vinyl—Class of resins containing materials such as polyvinyl chloride, polyvinyl acetate, etc.

Viscometer (*Viscosimeter*)—Instrument for measuring the viscosity of a fluid. Types of viscometers are: 1) Poiseuille, 2) Stokes, 3) Couette, 4) Newton, 5) Coulomb, according to the effect used to measure the viscosity.

Viscosity Breaking (*Visbreaking*)—Thermal treatment (cracking) of petroleum fractions, with or without catalyst.

Viscosity Gravity Constant—Correlation of specific gravity and viscosity: low for paraffinic, high for napththenic oils.

Viscosity Index—Relationship between viscosity change of an oil with change in temperature; an empirical measure of the temperature coefficient of viscosity.

Viscosity—Property of resistance offered by a fluid to relative motion of its parts.

Viscosity Range—Ratio of viscosity of a fluid at its lowest temperature to that at its highest temperature, or lowest to highest shear.

Viscous Fluid—Fluid having appreciable fluid friction.

Viscous Forces—Tangential frictional forces. The action of such internal shear forces results in a conversion of mechanical energy into heat or unavailable thermal energy.

Volatility—Relative tendency of a substance to evaporate. Controlled by vapor pressure for components in a homogeneous liquid obeying Raoult's Law.

Vortex Swirl—Rotation of a liquid as if it were a solid body; there is no interchange of position between particles.

Wax Distillates—Neutral petroleum distillates that contain paraffin.

Weber Number—Dimensionless parameter concerning surfaces or interfacional tension forces.

Wetting Agent—Surface-active agent that is widely used in the textile industry to aid dyeing and sizing operations by making the textile absorbent and thus easier to "wet" with the dye or size.

W/O Emulsions—Water-in-oil emulsion in which the water is the internal phase and the oil is the external or continuous phase. The W/O emulsion may diluted by the addition of an oil and retain its homogenuity.

Yield Strength—Lowest stress at which a material undergoes plastic deformation. Below this stress the material is elastic; above it, viscous.

Yield Value of Plastic—Minimum stress required to start flow.

Zeitfuchs Crossarm Viscometer—Instrument used to determine kinematic viscosities of transparent or opaque Newtonian liquids in the 0.3-10,000 cS range.

Zeitfuchs Viscometer—Instrument used to determine kinematice viscosities of transparent Newtonian liquids in the 0.6-5,000 cS range.

Zeta Potential—Measure of the speed with which charged oil droplets travel under the influence of a unit electrical field.

Zwitter Ions—Molecular units containing groups that are ionized cationically and groups that are ionized anionically, both present in approximately the same numbers so that the molecule as a whole is electrically neutral.

INDEX